IS SCIENCE SEXIST?

A PALLAS PAPERBACK / 18

MICHAEL RUSE

University of Guelph, Guelph, Ontario, Canada

IS SCIENCE SEXIST?

And Other Problems in the Biomedical Sciences

D. REIDEL PUBLISHING COMPANY

DORDRECHT : HOLLAND / BOSTON : U.S.A.

LONDON : ENGLAND

Library of Congress Cataloging in Publication Data

CIP

Ruse, Michael.
    Is science sexist?

    (The University of Western Ontario series in philosophy of science; v. 17)
    Bibliography: p.
    Includes indexes.
    1. Biology—Philosophy. 2. Sexism in sociobiology. 3. Sociobiology—
Philosophy. I. Title. II. Series.
QH331.R8787   303.4′83   81-8504
ISBN 90-277-1249-2   AACR2
ISBN 90-277-1250-6 (pbk.) (Pallas edition)

Published by D. Reidel Publishing Company,
P.O. Box 17, 3300 AA Dordrecht, Holland

Sold and distributed in the U.S.A. and Canada
by Kluwer Boston Inc.,
190 Old Derby Street, Hingham, MA 02043, U.S.A.

In all other countries, sold and distributed
by Kluwer Academic Publishers Group,
P.O. Box 322, 3300 AH Dordrecht, Holland

D. Reidel Publishing Company is a member of the Kluwer Group

*Also published in 1981 in hardbound edition by Reidel in the
University of Western Ontario Series in Philosophy of Science, Volume 17*

Printed in The Netherlands

For David Hull
With affection and respect

TABLE OF CONTENTS

ACKNOWLEDGEMENTS

I am obliged to the following for permission to draw on material written by me and published previously: to the Philosophy of Science Association, publishers of 'Karl Popper's philosophy of biology', *Phil. Sci.* **44** (1977), 638–61; and 'Sociobiology: Sound science or muddled metaphysics', *PSA 1976* **2** (1977), 48–73; to the Addison-Wesley Publishing Company, publishers of 'Are homosexuals sick', A. Caplan *et al.* (eds) *Concepts of Health and Disease* (1981). Additionally, earlier versions of two of the essays appeared in Reidel Publications: 'Reduction in genetics', A. C. Michalos *et al.* (eds) *PSA 1974*, pp. 633–52; and 'Genetics and the quality of life', *Social Indicators Research* **7** (1980), 419–41. The title of the whole collection (and Essay 9) was used by me in a short discussion note in *The Behavioral and Brain Sciences* **3** (1980), 197–8, and is used here with permission of the editor.

I am obliged also to the following for permission to use illustrations from books published by them: Figures 1.1; 2.1; 2.11; 2.12; 5.1; 5.2; 5.3; Tables 2.1; 5.1 from F. J. Ayala and J. W. Valentine, *Evolving: The Theory and Processes of Organic Evolution*, Menlo Park, California: The Benjamin/Cummings Publishing Company, 1979. Fig. 1.2 from G. W. Burns, *The Science of Genetics* 2nd edn, New York: Macmillan Publishing Co., 1972. Fig. 1.3 from M. Ruse, *The Philosophy of Biology*, London: Hutchinson, 1973. Fig. 1.4 from D. Lack, *Darwin's Finches*, Cambridge: University of Cambridge Press, 1947. Fig. 2.2 from the estate of the late Dr H. B. D. Kettlewell. Figs. 2.3; 2.4; 2.10; 2.15 from Th. Dobzhansky *et al., Evolution*, San Francisco: W. H. Freeman, 1977. Fig. 2.5 from The University of Chicago Press (publishers of J. E. Duerden, 'Inheritance of calosities in the ostrich', *Am. Nat.* **54** (1920), 289, the original source of the picture). Fig. 2.6 from M. Ruse, *The Darwinian Revolution: Science Red in Tooth and Claw*, Chicago: University of Chicago Press, 1979. Fig. 2.14 from J. M. Weller, *The Course of Evolution*, New York: McGraw-Hill, 1969. Fig. 3.1 from D. E. Schreiber, formerly IBM Research Laboratories. Fig. 3.2 from M. F. Boyd, *Malariology*, Philadelphia: Saunders, 1949. Fig. 4.1 from A. E. Brown and H. A. Jeffcott, *Absolutely Mad Inventions*, New York: Dover, 1970. Figs. 4.2; 7.6; 7.7 from the W. H. Freeman Company, San Francisco (publishers of *Scientific American* in which these illustrations first appeared). Fig. 4.3 from *The Canadian Entomologist* and Dr C. S. Holling, University of British Columbia. Fig. 5.4 from I. Lakatos and A. Musgrave, *Criticism and the Growth of Knowledge*, Cambridge: Cambridge University Press, 1970. Fig. 5.5 from G. R. Fraser and O. Mayo, *Textbook of Human Genetics*, Oxford: Blackwells, 1975. Figs. 5.7 and 6.1 from N. Rothwell, *Understanding Genetics*, New York: Oxford University Press, 1979. Fig. 5.8 from N. R. Hanson, *Patterns of Discovery*, Cambridge: Cambridge University Press, 1958. Fig. 5.9 from R. C. Lewontin, *The Genetic Basis of Evolutionary Change*, New York: Columbia University Press, 1974. Fig. 6.2 from J. Scrimgeour, *Towards the Prevention of Fetal Malformation*, Edinburgh: Edinburgh University Press, 1978. Fig. 6.3 from *The American Scientist*. Figs. 7.1; 7.2; 7.3 and 7.4 from D. Jackson and S. Stich, *The Recombinant DNA Debate*, Englewood Cliffs: *Prentice-Hall*, 1979. Figs. 7.5 and 7.8 from C. Grobstein, *A Double Image of the Double Helix*, San Francisco: W. H. Freeman, 1979. Fig. 9.1 from E. O. Wilson, *Sociobiology: The New Synthesis*, Cambridge, Mass.: Harvard University Press, 1975. Fig. 9.2 from D. Symons, *The Evolution of Human Sexuality*, New York: Oxford University Press, 1979, and Dr L. B. Shettles, Randolph, Vermont. Figs. 10.1 and 10.2 from G. Dörner, *Hormones and Brain Differentiation*, Amsterdam: Elsevier, 1976.

Finally, for permission to quote from their publications, I am obliged to: H. M. Morris (ed), *Scientific Creationism*, San Diego: Creation-Life Publishers. D. Symons, *The Evolution of Human Sexuality*, New York: Oxford University Press, 1979. E. O. Wilson, *On Human Nature*, Cambridge, Mass.: Harvard University Press, 1978.

The most consequential change in Man's view of the world, of living nature and of Himself came with the introduction, over a period of some 100 years beginning only in the 18th century, of the idea of change itself, of change over periods of time: in a word, of evolution.

Ernst Mayr

Well, it is a theory, it is a scientific theory only, and it has in recent years been challenged in the world of science and is not yet believed in the scientific community to be as infallible as it once was believed. But if it was going to be taught in the schools, then I think that also the biblical theory of creation, which is not a theory but the biblical story of creation, should also be taught.

Ronald Reagan

PREFACE

Philosophy of biology has a long and honourable history. Indeed, like most of the great intellectual achievements of the Western World, it goes back to the Greeks. However, until recently in this century, it was sadly neglected. With a few noteworthy exceptions, someone wishing to delve into the subject had to choose between extremes of insipid vitalism on the one hand, and sterile formalizations of the most elementary biological principles on the other. Whilst philosophy of physics pushed confidently ahead, the philosophy of biology languished. In the past decade, however, things have changed dramatically. A number of energetic and thoughtful young philosophers have made real efforts to master the outlines and details of contemporary biology. They have shown that many stimulating problems emerge when analytic skills are turned towards the life-sciences, particularly if one does not feel constrained to stay only with theoretical parts of biology, but can range over to more medical parts of the spectrum. At the same time, biology itself has had one of the most fruitful yet turbulent periods in its whole history, and more and more biologists have grown to see that many of the problems they face take them beyond the narrow confines of empirical science: a broader perspective is needed. Thus we find that philosophers have looked towards biology and biologists have looked towards philosophy, and at the meeting point an excitingly reinvigorated discipline has put down fresh roots and sent forth new branches, if one might be permitted a happy use of metaphor.

This book represents fruits gathered by one who has tended the plant (!) I offer the reader a collection of essays, some entirely new, some heavily revised versions of work which has appeared previously, which show the kinds of problems that have engaged one particular philosopher of biology. I have deliberately chosen and modified my work in such a way that, hopefully, the collection will read in a more uniform and integrated fashion than is usually the case with collections; also, believing that what I have to say might be of interest outside of the narrow professional philosophical community, I have tried to avoid specialized, technical discussion. Because the achieving of such unity and clarity have been major principles of selection, although I have included nothing by which I no longer stand, necessarily I have felt obliged to omit some work of which (immodestly) I am still somewhat proud.

Nevertheless, I think I can truly say that the collection gives a roughly chronological fair picture of my decade-long interests in the philosophy of biology, more broadly, the philosophy of the biomedical sciences. Moreover, I think it is probably true to say also that many of my interests have been typical of philosophers of biology as a whole, although, of course, this is not to say that others would accept the conclusions I draw.

The first two essays in the collection, Essay 1: 'The structure of evolutionary theory', and Essay 2: 'The evidence for evolutionary theory', set the scene for much that is to come later. Indeed, I try to introduce as much scientific content as possible into these essays, in order that the non-biological reader will have firm background for many of the discussions which follow. But in their own right, the essays try to do something which it seems to me is terribly important: establish for once and for all that the modern theory of evolution, the so-called 'synthetic theory' or 'neo-Darwinism', has every right to be considered a genuine, well-established theory. This noble product of the human mind is hounded on all sides, by well-known philosophers and by religious fanatics. The reader may not accept my particular philosophy of science, logical empiricism – indeed, some of the most brilliant recent work in the philosophy of biology has been directed at showing that evolutionary theory, like other great scientific theories, better exemplifies a revisionist philosophy of science – but I argue here and now, as I have always argued, that the synthetic theory is authentic science of the best kind.

In many respects Essay 3: 'Karl Popper and evolutionary biology' continues my defense of the synthetic theory. I examine the excursion into biology by the deservedly highly respected philosopher, Karl Popper. I argue that his many critical comments about evolutionary thought are ill-founded, based on a gross misapprehension of real biological theorizing. I show that judged by his own 'criterion of demarcation', falsifiability, the synthetic theory is a genuine theory, and that when Popper argues, as so many others have done, that the central evolutionary mechanism of natural selection is tautological, he is just plain wrong.

Essay 4: 'The last word on teleology, or optimality models vindicated', takes up a subject much-discussed by philosophers and biologists, namely whether there is something distinctively end-directed or 'teleological' about biological thought. I argue that there is; but at the same time I am at pains to point out that there is nothing other-worldly or flabby-minded about biologists. When they are compared to physical scientists they rate well. I show that any differences between physical modes of thought and biological modes of thought are a function of the material, not of the thinkers.

Essay 5: 'The molecular evolution in genetics', rounds out the first half of the collection. There has been much debate recently about the true relationship between the older theory of heredity, Mendelian genetics, and the glory of mid-twentieth-century biology, molecular genetics. Does the older theory get entirely superceded by the newer theory, or is the older in some sense taken up and incorporated into the newer? I try to show that the incorporation thesis, or 'reduction' as it is better known, has much to commend it and that, contrary to the claims of at least one well-known philosopher, such a perspective makes sense of much that biologists do and claim.

We come next to essays dealing with topics where moral concerns are mingled more and more intimately and overtly with biological theorizing. In Essay 6: 'Does genetic counselling really raise the quality of life?' I take a hard and not entirely sympathetic look at the ways in which biomedical professionals are attempting to reduce the incidences of so-called 'genetic diseases', through such techniques as examination of a fetus's gene-structure by amniocentesis, with the possible option of an abortion of a defective fetus. Without wanting to be absolutely negative, I suggest that, like so many advances of modern medicine, the benefits might well be outweighed by the costs. At least, I plead that we should not get carried away uncritically by technological innovations.

Then in Essay 7: 'The recombinant DNA debate: a tempest in a test tube?' I take an essentially retrospective look at a controversy which raged bitterly for a while inthe 1970s: the controversy over the insertion of the genetic material of one organism into the genetic material of another organism. I argue that there are some fascinating lessons here, in the 'recombinant DNA debate', particularly about who is the proper authority for decisions affecting the safety and well-being of us all. Also, I try to mount a defense of science simply considered as science. All too often, the only reasons given for doing science centre on hypothesized technological benefits. I argue that good science in itself is as worthwhile a human achievement as the writing of a great opera or the formulating of a seminal philosophical system. For this reason we should cherish it.

The next two essays carry us into an area where biologist is pitted against social scientist and biologist is pitted against philosopher. The new area of the life sciences known as 'sociobiology', the attempt to understand animal social behaviour from an evolutionary perspective, has been in critical hot water from its beginning. Not content with applying their models to such neutral organisms as insects, the sociobiologists have moved right in and argued that their explanatory models can be applied illuminatingly to our own species,

Homo sapiens. In Essay 8: 'Sociobiology: sound science or muddled meta-physics?' I look at some of the general sociobiological claims, animal and human, particularly as they are to be found in E. O. Wilson's, *Sociobiology: The New Synthesis*. I defend the sociobiologists against philosophical criti-cisms, although I stop short of outright endorsement. Then in Essay 9: 'Is science sexist? The case of sociobiology', the title-essay of this collection, I consider a major charge that has been brought against sociobiology: that it is sexist, implying implicitly and explicitly that males are superior to females. I concede that it is certainly possible for a science to be sexist, Freudian psychoanalytic theory may well be a real example; but I argue that, to date, the charge against human sociobiology does not succeed. My defense is based on the thesis that the sociobiology of sexuality could be true.

Finally in this collection, in Essay 10: 'Are homosexuals sick?', I draw on recent philosophical analyses of the concepts of 'health', 'disease' and 'illness', in order to throw light on a matter of great individual and social importance. By comparing the philosophical analyses with recent empirical findings together with various putative etiologies for homosexual orientation, some of which have been introduced already in earlier essays, a number of different conclusions to the title-question can be obtained. I suggest that here we find the reason why there is so much disagreement about the health-status of homosexuals; but I suggest also that when we compare all the conclusions one fairly concrete answer to the question starts to emerge. Given the hatred, and bigotry, and condescention, that variant sexual orientation causes in our society, I shall feel justified in what I have written if I can persuade but one person to change their future beliefs and actions towards the course of reason and away from that of emotion.

In concluding this preface, having told the reader what I intend to do, let me pay full credit to those who have helped me to do it. I am bound to miss some who deserve mention, but at least let me name: John Beatty; Christopher Boorse; Mario Bunge; Arthur Cain; Arthur Caplan; Lindley Darden; Thomas Goudge; Ernst Mayr; Alex Michalos; John Richards; Ken-neth Schaffner; Paul Thompson; Mary Williams; William Wimsatt. I count myself lucky to have so many professional colleagues who are also good friends.

As always, when I put pen to paper, I sense the friendly, but critical eye of David Hull looking over my shoulder. I speak now for myself, but I know I speak for all, when I say that I cannot thank him enough for all that he has done. Even when we disagree, and perhaps partially to prove to myself that I am not too far under his shadow, I have included in this collection an essay

where we do disagree, I can truly say that he is "an opponent, whose intellectual and moral fairness it is a pleasure to acknowledge".

For the past ten years, Judy Martin has typed, re-typed, and re-retyped.
She combines unfailing good humour with the highest professional standards.
Towards the end she has been ably assisted by Marilyn Watt and Vivien Keir.

THE STRUCTURE OF EVOLUTIONARY THEORY

In 1859 Charles Robert Darwin, well-known Victorian traveller, authority on barnacles, and long-time recluse, published his epoch-making work: *On the Origin of Species by Means of Natural Selection; or The Preservation of Favoured Races in the Struggle for Life*. Neither Darwin nor biology were ever quite the same again. Denounced by bishops, defended by scientists, satirized by novelists, Darwin's major claim — that the whole organic world, right up to and including ourselves, is the product of a slow, natural, 'evolutionary' process — rapidly became one of the most controversial issues of the day (Ruse, 1979a). And yet, Darwin and his ideas stood their ground, gradually winning support. From being heresy they became, if not orthodoxy, almost respectable. When Darwin died in 1882, less than a quarter-century after the *Origin* appeared, he was carried to his grave by two dukes and an earl, past, present and future presidents of the Royal Society, and the American Minister (F. Darwin, 1887). Significantly and appropriately, he rests in that English Valhalla, Westminster Abbey, but a few feet from Isaac Newton.

One man can only do so much. Copernicus may have put the sun at the centre of the universe, but the break with circular celestial motion had to wait for Kepler (Kuhn, 1957). Similarly, for all of his advances, Darwin's own contribution to his revolution was limited and incomplete (Eiseley, 1958). Darwin saw and defended organic evolution. He saw also that many more plants and animals are born than can possibly survive and reproduce on this small planet of ours. Hence, there is bound to be a 'struggle for existence', and success in the struggle will lead to a constant winnowing of organisms. Darwin realized that his process, 'natural selection', could have profound evolutionary effects. But he could not really see how characteristics get passed on from one generation to the next — why big pigs give birth to little pigs rather than to cabbages, and why big black pigs tend to give birth to little black pigs rather than to little pink pigs — nor even more importantly, could Darwin see how we get new organic variations, the 'raw stuff' of evolution. For this information, the world had to wait until this century, which brought the rediscovery and development of the ideas of Darwin's contemporary, the obscure Moravian monk, Gregor Mendel, who worked out the essential principles of heredity in isolation in his cloistered garden in Brno (Olby, 1967).

In love and science nothing ever runs very smoothly, and it will come as no surprise to learn that when the newly rediscovered Mendelian 'genetics' was being articulated and expanded, many thought that it and Darwinian natural selection were rival evolutionary mechanisms: one could not be both a Mendelian and a Darwinian at the same time (Provine, 1971). It was not until around 1930 that it was realized that Darwinism and Mendelianism complement, rather than contradict, each other, both providing essential parts of the overall evolutionary picture. Happily, following this realization, theoretical, experimental, and observational work rapidly fleshed out the Darwin–Mendel hybrid, until it grew into what has come to be known as the 'synthetic theory of evolution'. In 1937, Theodosius Dobzhansky produced his *Genetics and the Origin of Species*, in which he presented the fruits of his early experimental studies of evolutionary processes, particularly as they can be discerned in Drosophila (fruitflies). Then, in 1942, Ernst Mayr in *Systematics and the Origin of Species* discussed evolution in the animal world around us today and in 1944, George Gaylord Simpson did the same for the past in *Tempo and Mode in Evolution* (revised and retitled edition, Simpson, 1953). G. Ledyard Stebbins did for flora, in 1950 in his *Variation and Evolution in Plants*, what others had done for fauna. And at the same time in Europe as well as America, these and other aspects of the evolutionary picture were articulated and extended (see, for instance, Huxley, 1942).

In Thomas Kuhn's well-known language, in the Darwin–Mendel synthesis, evolutionary biology had got its 'paradigm': "[An] achievement ... sufficiently unprecedented to attract an enduring group of adherents away from competing modes of scientific activity. Simultaneously ... sufficiently open-ended to leave all sorts of problems for the redefined group of practitioners to resolve" (Kuhn, 1970, p. 10). Moreover, I think it is true to say that, although in the past thirty years or so there have been significant advances on the evolutionary scene, essentially it is the synthetic theory of evolution within which today's evolutionists work (Dobzhansky *et al.*, 1977; Ayala and Valentine, 1979). Certainly, it is this theory which provides evolutionists with their background assumptions and the problems which need solving, and it is into this theory that students new to the field are initiated. Admittedly, like all vigorous areas of science there are hotly debated topics within the theory. As is so often the case when siblings quarrel, one detects a note of bitterness that would be absent were the disputants farther apart. But, generally speaking, there is much agreement about the fact of evolution and about the synthetic theory's proposals for explaining this fact.[1]

Nevertheless, the synthetic theory, or 'neo-Darwinism' as it is sometimes called, continues to attract suspicion and hostility from outside. Few today dispute (say) Einstein's theory of relativity, for all its initial implausibility. However, in serious journals, in popular magazines, and even in newspapers, discussions critical of the theory of evolution erupt periodically. We are told that it is 'just a theory', or 'only a truism', or (somewhat inconsistently) 'downright false', or some such thing. More charitably we learn that neo-Darwinism is a hypothesis; less charitably we learn that it is a speculation (Bethell, 1976; Macbeth, 1971; Popper, 1972, 1974a; Koestler, 1971). Some of the opposition is easy to understand, and is, I am afraid, all-too-easy to underestimate. There are probably few today who support a geocentric world for religious reasons, despite the sun's supposed stopping for Joshua, but there are still those who feel their Christianity threatened by any kind of evolutionary hypothesis or earth-span of much more than the traditional 6000 years. (Whitcomb and Morris, 1961; Gerlovich et al., 1980). For people like this, the synthetic theory is anathema and they have attacked it vigorously; in many cases with success sufficient to keep conservative religious views as part of school curricula and to keep evolutionary views muted in high-school texts. But there are others, more 'respectable' thinkers, who also are troubled by the synthetic theory. Many suspect that the theory's success, if success it be, comes only by default — the synthetic theory triumphs simply because no one has put forward a convincing alternative (Manser, 1965; Grene, 1958; Himmelfarb, 1962). And even those who concede that perhaps evolutionists are thinking roughly along the right lines, are appalled at what they see as the characteristic flabbiness of the theory (Scriven, 1959; Smart, 1963). Somehow, there is a feeling that, judged by the standards of the best theories of science, that is to say, the theories of physics and chemistry, the synthetic theory of evolution is a dreadful disappointment.

I think that all of these critics of modern evolutionary theory, whether they be religiously inspired or not, are wrong. In the next essay I shall take up the question of the evidence for the theory, concentrating in this essay on the preliminary question of the actual nature and structure of the synthetic theory. I admit to a certain looseness to the theory — almost to be expected in something as new, conparatively speaking, as evolutionary theory, and certainly to be expected in something dealing with such difficult material as does evolutionary theory. But this concession notwithstanding, I argue simply that in crucial respects the synthetic theory of evolution, neo-Darwinism, is very much like a theory in the physical sciences. Putting the matter bluntly, and at the moment shelving the question of evidence, I argue that the synthetic theory of evolution is a genuine scientific theory. Please note that it is not my claim that in no respects does evolutionary theorizing differ from the theorizing to be found in the physical sciences. Indeed, I think there is a crucial difference between physics and biology, and, in a later essay, I shall explore and analyse this difference in some detail. But there are major similarities which many overlook or ignore, and it is on these that my discussion in this essay devolves. Parenthetically, let me add that, because I shall argue that there are these similarities between the synthetic theory and physico-chemical theories, I shall ignore entirely the rather tiresome side-question as to whether there could, in fact, be genuine scientific theories which are not in any way like those of physics and chemistry.

1.1. THREE FEATURES OF PHYSICO-CHEMICAL THEORIES

Of course, the claim that the synthetic theory is like a theory in the physical sciences presupposes that we know the nature of a theory in the physical sciences. Unfortunately, it is simply impossible to make any non-controverted claim about physico-chemical theories, so in order to get to my main theme, I will simply state what I believe to be the case, sounding somewhat more dogmatic and self-confident than I really feel (Ruse, 1976a). I suspect strongly that although I risk alienating good and respected friends in the philosophical world with my stand, many of the points I want to make about biology would go through were one to adopt a different philosophy of physics than that to which I subscribe. And whilst I have an instinctive distrust of majorities, for once I can take consolation in numbers, for I believe my position is the common one. Even its critics refer depreciatingly to it as the 'received view'. (See Hempel (1965, 1966), and Nagel (1961) for general statements of this position, and see Suppe (1974), Beatty (1978, 1980) and Giere (1979) for pertinent criticisms.)

Looking at the great theories of physics — Newtonian astronomy, the wave theory of light, quantum mechanics — a number of features stand out, from which I select the following three. First, we get *reference to entities of different kinds*, or at least to entities which we can approach only in different ways (Ruse, 1973a). We ourselves live in a comfortable, familiar world of chairs and tables, and (if we are scientists) of prisms and pendulums and planets, and so forth. All of these are things which, one way or another, we can see or otherwise sense. And physical theories talk about these sorts of things. Kepler's laws, for instance, tell us about what the planets do. Wave theory tells us about what sorts of light and shadow patterns we might expect as, for instance, in Young's double slit experiment. But, also in physical theories, we find reference to other kinds of things — things which are not visible or tangible. We get told about gravitational attractions, and light waves, and electrons — things which in some way we have to infer, because they are not directly evident to our senses. They are 'theoretical', meaning that they come to us through (and only through) the theory, and often their existence is (or was) in some doubt — they are hypothesized. It is these non-observational or theoretical entities, as I will now refer to them, which give our theories their explanatory power. We explain the planets in terms of attractions and forces, not *vice versa* and, similarly, we explain patterns of light and dark in terms of light waves. It is not being claimed that the distinction between observational and theoretical entities is absolute nor that there are no borderline cases, but it does seem to be there nevertheless.

Second, physico-chemical theories are axiomatic or *hypothetico-deductive*, in the sense that one set of claims can be seen to follow deductively from another set. About this, Richard B. Braithwaite writes as follows:

A scientific system consists of a set of hypotheses which form a deductive system; that is, which is arranged in such a way that from some of the hypotheses as premisses all the other hypotheses logically follow. The propositions in a deductive system may be considered as being arranged in an order of levels, the hypotheses at the highest level being those which occur only as premisses in the system, those at the lowest level being those which occur only as conclusions in the system, and those at intermediate levels being those which occur as conclusions of deductions from higher-level hypotheses and which serve as premises for deductions to lower-level hypotheses. (Braithwaite 1953, p. 12.)

It should be noted that, in the physical sciences, unlike *a priori* enterprises like pure mathematics and logic, although all the general statements of the systems are presumed true within their limits, not all are thought true simply by virtue of their form. A claim like (say) Boyle's law or Snell's law is empirical — that is to say, its truth or falsity is a function of the way that the world actually is. However, it is believed that the general claims of science transcend the merely contingent: it is thought that given the world as it is, they *must* hold. This sense of necessity is often referred to as 'nomic' necessity, and the statements of science are said to be 'lawlike' (Nagel, 1961). Perhaps God could have made a world where Boyle's law fails, but given this world as it is, one simply cannot have Boyle's law breaking down, so long as we are working with a gas within the usual limits.

Referring back to our first point, what we find is that the upper-level hypotheses of the theory contain reference to the theoretical entities, and then through the deductive process these hypotheses explain the lower-level statements of the theory, which refer to observational entities, the objects of our experience. Explanation, therefore, is a question of showing that something follows deductively from something else. Obviously, since the upper-level claims talk of theoretical entities and the lower-level claims of observational entities, at some middle-point in theories we need bridge or translation principles to take us from talk of one kind to talk of the other kind. In gas theory, for instance, we are told that we can relate presure (observational) with little particles striking against a wall (theoretical), and temperature (observational) with the speed of such particles (theoretical) (Holton and Roller, 1958).

Third, the best physical theories are what the philosopher William Whewell (1840) referred to as *consilient*. They do not explain in just one area, but draw together disparate parts of their science, showing how all are intimately connected and follow from one or a few over-arching principles. Thus, in Newtonian mechanics, we get both Galilean terrestrial dynamics and Kepler's laws of the planets, not to mention many other parts of physics, brought together and bound beneath the same laws of motion and the same law of gravitational attraction. In physics and chemistry we look for economy and simplicity, as hitherto-unconnected parts are made into a unified whole. Incidentally, if I might be permitted to make a point directed towards those biologists who feel a little upset with the fact that I would even dare to compare biology with physics, I would add that this consilience is a major factor in other areas of science (Ruse, 1976b, 1980a). Recently, supporters of the new plate tectonics in geology wrote as follows:

Certainly the most important factor is that the new global tectonics seem capable of drawing together the observations of seismology and observations of a host of other fields, such as geomagnetism, marine geology, geochemistry, gravity, and various branches of land geology, under a single unifying concept. Such a step is of utmost importance to the earth sciences and will surely mark the beginning of a new era. (Isacks *et al.*, 1968, p. 362.)

Returning to the main thread, there are other aspects of physical theories that one could highlight. Nevertheless, these points just given are crucial and representative, and suffice for our purpose. I would not claim that absolutely every good theory of physics and chemistry shows everything in every respect − often the axiomatic nature of theories is rather sketchy. However, I do not think it waters down the position just presented too far to say that the 'received view' acts as something of an ideal or model for physical scientists − for various reasons they may not always satisfy exactly every point, but the various points do act as guidelines or signposts towards what the physicist or chemist aims to produce.

Turning now to the synthetic theory of evolution, my claim simply is that this theory fares well when considered against this picture of science, even though the picture was derived from the physical sciences, Hence, there is good reason to count the synthetic theory, neo-Darwinism, as real science. Perhaps the theory is immature in some respects; but judged against the criteria given it seems genuine for all that. Or perhaps more rigorously we should say that the theory is a proper candidate for genuine science. Remember, at this time I am not too concerned about evidence.

Of course, in comparing the synthetic theory against physico-chemical theories, we have to have before us some conception of what the synthetic theory of evolution is all about. I suppose to most people talk of 'evolution' summons up all sorts of images of fossils and dinosaurs and the bits and pieces of bone that the Leakeys are digging up in Africa (Isaac and McCown, 1976; Leakey and Lewin, 1977; Washburn, 1978). However, although things like these are undoubtedly part of the overall evolutionary picture, one of the consequences I shall draw out from my discussion in this essay is that starting with these in introducing the synthetic theory would be to start at the wrong end of the theory. Where we must begin, rather, is down at the level of the individual — today's individual. In short, where we must begin is with that part of biology which deals with organic characteristics, their transmission, their appearance and disappearance. We must start with genetics. From those of my readers who feel that there is something suspiciously misleading about all of this, can I simply beg tolerance for a short period of time? I shall speak to their concerns later. At the moment, if an argument from authority carries any weight, let me merely note that if one turns to the classics of the synthetic theory, Dobzhansky's *Genetics and the Origin of Species* or Simpson's paleontological works, for instance, genetics seem to figure prominently in the early pages! Thus, the way I go about things may not be quite so wrong after all. (But for rather different philosophical approaches to mine, see Williams, (1970) and Rosenberg (1980).)

1.2. EVOLUTIONARY THEORY AND THE OBSERVATIONAL / THEORETICAL DICHOTOMY

Enough of apologies. Let us turn to biology. Instead of presenting the whole of the synthetic theory and then, if I have any sufficiently patient readers left, analysing the theory with respect to the three points of comparison, I shall offer exposition and critical analysis simultaneously. This being so, I ask the first of our questions, namely that about the kinds of entities or things to which the synthetic theory refers. Do we see a distinction between theoretical or non-observational entities and non-theoretical or observational entities, where in some sense the former are being used to provide a causal explanation of the latter? I would suggest very strongly that we do, especially if we consider what I take to be the starting point for modern evolutionary thought, namely genetics. (I am about to give a very brief exposition, omitting all difficulties and exceptions. One really ought to consult one of the many, excellent, standard texts of genetics, for instance, Strickberger (1975).)

For convenience restricting ourselves to sexual organisms, the simplest way to begin is with the beginning of life for an organism when we have a fertilized egg, the 'zygote'. The zygote is a self-contained biological unit, a 'cell', and essentially the whole growth and life of the organism reduces to the duplication and reduplication of this initial cell, obviously with much specialization as the different parts of the body with different functions are formed. The key question, of course, is what 'programmes' this growth, carrying the information necessary to produce all the body's various features — hearts, eyes, hair, feet, and even types of instinctive behaviour? It is believed that the ultimate unit is the *gene*, or rather many genes, which are carried on long thread-like bodies ('chromosomes') in the centres of cells. The genes replicate themselves, thus ensuring that their information is passed on to newly formed cells and then, in various ways, the genes get triggered into action, bringing about the different products which are required for the different parts of the body. Thus, the totality of the body's physical characteristics, its 'phenotype', is in some ultimate fashion, causally dependent on the totality of the body's genes, its 'genotype'. Biologists are quick to add that, truly speaking, it is the genotype, in interaction with the external environment, which gives the full causal background to the development of organisms.

Taking up the story of the genes in a little more detail, we find that normally the chromosomes occur in pairs, and thus any particular gene on a particular place on a chromosome (a 'locus') has a corresponding mate on the corresponding chromosome. In a population of like organisms, it may be that, considering genes which can occupy any particular locus, a number of variants occur: these are known as 'alleles'. If, at some particular locus, an organism has identical alleles, it is said to be 'homozygous' (it is a 'homozygote'); if the alleles are different, it is 'heterozygous' ('heterozygote'). An organism may be homozygous with respect to one locus and heterozygous with respect to another locus. The products of heterozygous genes might duplicate one of the homozygote pairs, they might make for intermediates, or one might get quite new and different products. Where the heterozygote is like of the homozygotes, the allele of this homozygote is said to be 'dominant' over the allele of the other homozygote, which latter is said to be 'recessive'.

So far I have concentrated on the gene as a unit of *function*. It is also a unit of *heredity*, in the sense that it is genes which organisms pass on from one generation to the next, thus causing a continuity of physical characteristics (Hull, 1974). In sexual organisms each parent provides half the chromosome set (and thus, half the genes) of any new offspring, this being done through special sex cells (gametes) which unite to form new zygotes (Figure 1.1). Normally, the genes (or rather, copies) go on from generation to generation, in new combinations certainly, but essentially entire and unchanged in themselves. Mendelian genes are 'particulate'. However, sometimes spontaneously as it were, genes do change, and these changes can give rise to new products and, thus, to changes in phenotypes. This phenomenon, 'mutation' (giving rise to 'mutants'), is random in the sense that it cannot be predicted for any particular organism and also in the sense that it does not produce physical characteristics which are necessarily related to an organism's particular needs — in fact, most mutations are deleterious in that their possessors are worse off with them than they would be without them. However, on average, most mutations do occur with sufficient regularity that their rates can be quantified, and, in some cases, mutations are of value to their possessors. Thus mutations are the point at which evolution ultimately begins, in that it is they that provide the building blocks for the evolution of us all from blobs of jelly in the sea. Of course, inability to predict particular mutations does not imply that they are uncaused and, in fact, much progress has now been made in uncovering causes of mutation (Watson, 1975).

Returning to the first point of comparison between biology and the physical sciences, I suggest that in the just-described geneotype/phenotype dichotomy we have a paradigmatic example of the theoretical/observational dichotomy and, thus, in this crucial respect, evolutionary biology is like physical science. The gene is not directly visible, but rather is something inferred and is understandable and approachable only through a body of theory. We do not see genes, or touch them, or taste them. Instead we posit their existence, arguing that they give rise causally to certain effects. They function, therefore, as ultimate units, much like light-waves or molecules. They are invoked to explain phenotypes, which do indeed exist at the level of the senses. We see the stripes on the tiger. We smell the odour of the skunk. We hear the howling of the dog. These and like phenomena are all

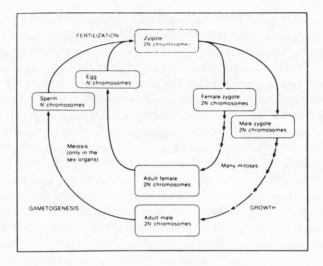

Fig. 1.1.   "The cycle of growth and reproduction. Most organisms develop from zygotes having two sets of chromosomes (2N), one inherited from each parent. The zygote divides into two cells, and each one of these into two others, and so on, by means of a process known as mitosis, which ensures that each cell in the body has the same two sets of chromosomes as the zygote. The sex cells are an exception; by means of a process, known as meiosis, two cell divisions take place with only one duplication of the chromosomes, thus sex cells have only one set of chromosomes (N). When fertilization takes place two sex cells, each with one set of chromosomes, unite and thus produce a zygote with 2N chromosomes," (From Ayala and Valentine, 1979, p. 29.).

A cell with one set of chromosomes only is said to be 'haploid', with two sets of chromosomes, 'diploid'. It must be emphasized that we are speaking here of the norm: some organisms are haploid throughout. The difference between male and female is a function of the so-called 'sex chromosomes'. In mammals, females have two (complementary) large chromosomes, the 'X' chromosomes. Males have an 'X' chromosome and a smaller 'Y' chromosome; thus females are 'homogametic' and males are 'heterogametic'. In some other organisms, for instance birds, it is males which are the homogametic sex and females the heterogametic sex.

observational in the way that prisms and pendulums are. In short, we find the two-level existence in evolutionary theory that we find in physics — or, if you prefer, we find the two-level way of knowing about existence. Moreover, there are the expected bridge principles. We are told what genotypes to associate with what phenotypes and whether, for instance, one allele is dominant over another, and whether, as a consequence, the heterozygote looks like one of the homozygotes or not. I shall give specific examples shortly.

Before we move on to consider the second point of comparison between evolutionary theory and physical theory, an objection must be faced and answered. Am I not making unduly heavy weather of matters at this point — in my eagerness to score philosophical triumphs, do I not ignore the history of science? The question at issue is whether or not evolutionary biology is like the physical sciences in showing a theoretical/observational dichotomy. The model is the sort of distinction we draw between molecules and gross physical characteristics. But surely we know now that biologists draw this distinction, not because genes are like molecules, but because they *are* molecules! Since the molecular revolution, most particularly since the pioneering work of Watson and Crick (1953), we know that the traditional biological unit, the gene, is in fact, a long piece of macromolecule, deoxyribonucleic acid (DNA). Thus, it is DNA which is responsible for phenotypes and thus, the thesis that biology is like physics is trivially true because biology is physics (Smart, 1963).

In reply, let me begin by admitting to the existence and importance of the molecular revolution. I certainly do not want to conceal it and shall, indeed, have things to say about it in a later essay in this book. However, on the one hand, in my own defence let me say that, even if we do identify the gene with DNA (which I do), it does not seem to me that this automatically makes evolutionary biology part of physics. In talking of DNA we are still concerned with the organic world, which is after all the subject of biology. Hence, even if we can make the case for the theoretical/observational dichotomy in evolutionary thought more easily than we suspected, the case is worth making all the same. On the other hand, let me note that, in many respects, the molecular revolution has not made all that much difference to evolutionary studies. Although the DNA model is introduced into discussions at an early point, evolutionists revert rapidly to talk about 'genes', and in most of their explanations it is the unanalysed biological entity which is the causal unit, not some physico-chemical import. (See, for instance, Dobzhansky (1970).) Do not misunderstand me. I do not argue that molecular biology has had no effect on evolutionary biology. It has had crucially important effects, some of which will be discussed later. Nevertheless, I do claim that the biological gene (whatever it 'really' is) thrives in biological discussions, and therefore it is still worthwhile establishing the point that, *vis-à-vis* this entity, one can make a strong case for the theoretical/observational dichotomy. And this I have tried to do. Without lingering further on a rather tangential point, let me note in passing that in one of the later essays of this book (Essay 5) I shall try to show just why it is that evolutionists can do so much of their work, almost as if the molecular revolution had never occurred. I shall suggest that this stems naturally from the logic of the situation, and it is not a reflection of biological obtuseness or physics phobia!

1.3. IS EVOLUTIONARY THEORY HYPOTHETICO-DEDUCTIVE?

We come to the next point of comparison between physical theory and evolutionary theory. Is the latter hypothetico-deductive, meaning that it is an axiom system deductively connecting general law-like statements, the upper levels referring to theoretical entities and the lower levels to observables? I think a case can be made for saying that the deductive model is indeed the ideal, although as admitted, the full theory is somewhat sketchy. Let us continue with our exposition, showing where the case can be made most strongly. The more evidence we can find that the synthetic theory does approach the axiomatic form in some parts, the more plausible is the claim that the axiomatic form functions as the ideal generally.

To begin, we have the fact that reference to the genes does not just occur haphazardly within the theory: rather they are introduced in the context of two general statements, Mendel's laws. (Let me not prejudge the issue by assuming that these are genuine laws.) These laws (or 'laws') dictate the way in which organisms pass on genes from one generation to the next. It has been noted already that, in sexual organisms, each parent contributes equally to the genotype of the offspring. The first law takes formal note of this fact, specifying also that there is a randomness about the particular contribution of the parents: "Given any sexual individual, each parent contributes one and only one of the alleles at any locus. These alleles come from the corresponding loci in the parents, and the chance of any parental allele being transmitted is the same as the chance of the other allele at the same locus." The second law specifies that the chances of any parental allele at one locus being transmitted are independent of the chances of an allele at a different locus. It is now known to hold only for loci on different chromosomes. (George (1964) has an exceptionally clear presentation of Mendel's laws, and I have modelled my presentation on hers.)

In this essay, I am deliberately trying to avoid too much discussion of justification and evidence, but in support of the right of Mendel's laws to be considered as genuine laws (i.e., as genuine as statements of physics), it can be asserted confidently that the general claims they embody have been found to hold as universally and in about as many diverse circumstances as any law of physics (Strickberger, 1975). And the very reason why we think we are dealing with a law of nature and not just a matter of contingent fact, is that law claims do, in fact, hold in the most diverse set of circumstances. They keep proving true time and again until we think they 'have to be' true (Ruse, 1970). Therefore, by criteria like this, Mendel's laws qualify as genuine laws. Just as Boyle's law holds for gas after gas, so Mendel's laws hold for organism after organism. They are true of everything, from cabbages to kings, whether these latter be human or penguin. Admittedly there are exceptions and restrictions, but no more than one finds in any law of physics. Snell's law does not hold for Iceland Spar; Boyle's law does not hold for high temperatures and pressures. Moreover, just as one can set up more refined versions of Boyle's law (e.g., van der Waal's equation), so also biologists have found ways of refining Mendel's laws. It was mentioned just above that Mendel's second law breaks down for loci on the same chromosome. It has been found however that this does not imply total chaos for genes on the same chromosome; rather such genes in turn are governed by strict rules, just as gases at non-standard temperatures and pressures are governed by strict rules. Thanks to a phenomenon known as 'crossing over', whereby paired chromosomes break at certain points, rejoin with the complementary chromosome (i.e., join up again 'crossed over' and, thus, exchange complete segments of themselves), genes on the same chromosome get shuffled during heredity, and this process can be quantified (Dunn, 1965; Carlson, 1966).

Moving along reasonably briskly we come next to what I believe is the lynch-pin of the whole evolutionary picture, the Hardy–Weinberg law or equation (Dobzhansky, 1951, 1970). This law states that if we have a large population (effectively infinite) which is mating at random and if, at a particular locus, we have two alleles (A_1 and A_2) in ratio $p : q$ (i.e., $p + q = 1$), then whatever the initial distribution, if there are no external disrupting forces, in the next and all succeeding generations the distribution of genotypes will be $p^2 A_1 A_1 + 2pq A_1 A_2 + q^2 A_2 A_2$ and, furthermore, the ratio of A_1 to A_2 alleles stays constant at $p : q$. ($A_1 A_1$ is an A_1 homozygote, etc.)

This law is not plucked out of thin air. It can be readily shown that it is a deductive consequence of Mendel's first law (Falconer, 1961; Li, 1955; Ayala and Valentine, 1979). In fact, one can draw up a little matrix feeding in all the given information, showing how the ratios follow:

		Gene frequences among males	
		$p\,(A_1)$	$q\,(A_2)$
Gene frequences	$p\,(A_1)$	$p^2\,(A_1 A_1)$	$pq\,(A_1 A_2)$
among females	$p\,(A_2)$	$pq\,(A_1 A_2)$	$q^2\,(A_2 A_2)$

These frequences obtain because (by Mendel's first law) the probability of getting a homozygous individual A_1A_1 is the probability of getting an A_1 from the father multiplied by the probability of getting such an A_1 from the mother, that is $p \times p = p^2$. Similarly, for the homozygote A_2A_2, $q \times q = q^2$. In the case of the heterozygote, we have $p \times q$ chance of an A_1A_2 individual and $q \times p$ chance of an A_2A_1 individual, and since these individuals are of the same kind, the total chance of a heterozygote is $2pq$. Finally, it is easy to show that the new gene ratio is $p : q$. We have p^2 A_1A_2 individuals, which yield $2p^2$ A_1 genes, and we have $2pq$ A_1A_2 individuals yielding $2pq$ A_1 genes (each homozygote has two A_1 genes, whereas heterozygotes have one A_1 gene). Thus, for A_1 genes, we have $2p^2 + 2pq = 2p(p+q) = 2p$. Similarly, for A_2 genes we have $2pq + 2q^2 = 2q(q+p) = 2q$. The ratio therefore is $2p : 2q$, or $p : q$. (It will not have escaped the notice of the discerning reader that, in order to obtain the Hardy—Weinberg law, we have to assume not simply that the ratio of A_1 to A_2 alleles is $p : q$, but that this ratio holds in *both* males and females. If all the males were A_1A_1 homozygotes and all the females A_2A_2 homozygotes, one could get nothing but A_1A_2 heterozygotes).

It is impossible to underestimate the importance of the Hardy—Weinberg law for evolutionary thought. It is sometimes suggested that it is basically little more than a truism — essentially it seems to say that if nothing happens (i.e., no external forces), then nothing happens (i.e., gene ratios stay the same). However, this is to misread the law's status and function. If Mendel's first law did not hold, then given two alleles there would be no barrier (say) to whichever allele was the more common eliminating the other allele totally in a few generations (Provine, 1971). What the law guarantees is the persistence of minorities, no less than majorities. In fact, the law functions very much as does Newton's first law of motion in his mechanics. This latter law, also an 'if nothing happens then nothing happens' kind of law, guarantees a background of stability. Objects do not speed up or slow down, without good cause. Similarly the Hardy—Weinberg law provides a background of genetic stability. Gene ratios do not alter around, without good reason (Ruse, 1973a).

The Hardy-Weinberg law is like Newton's law in another respect, namely that normally there are factors disrupting the equilibrium which the laws guarantee in the disruption's absence! However, sometimes there are no apparent disruptions and the expected equilibria do indeed obtain. For instance, one study of $M-N$ blood groups in the English found the following genotypic frequencies:

MM	*MN*	*NN*
28.38	49.57	22.05

The theoretical Hardy–Weinberg ratios are:

MM	*MN*	*NN*
28.265	49.800	21.935

To the untrained eye, this is a pretty good fit! Indeed, the researchers noted that even if one does have equilibrium, sampling errors lead one to expect such an exact fit less than one time in ten (Race and Sanger, 1954).

There are two principle threats to the genetic stability of a population, and it is at this point that we start to get to the crux of the evolutionary process. Already we have encountered one threatening factor. It is mutation, the source of new variation. The other potentially disruptive factor is Charles Darwin's special contribution to the evolutionary picture: *natural selection*. In the spirit of Darwin (1859) it is argued by evolutionists that the physical characteristics to which genes give rise can help or hinder their possessors in the struggle to survive and reproduce (Maynard Smith, 1975; Ayala and Valentine, 1979). Those characteristics which help are said to give their possessors the 'adaptive' edge, and are themselves spoken of as 'adaptations' (Williams, 1966). The organisms themselves and the genes, which are responsible, are said to be 'fitter' than other organisms and genes – they have superior 'fitness'. But, language apart, what this all means is that some genes, because of their effects, increase their representation in future generations at the expense of other genes. Thus, through this differential reproduction, as with mutation we have disruption of Hardy–Weinberg equilibrium. Mutation brings new genes into a population. Selection alters around the ratios of those genes that are there (Li, 1955).

One fact must be acknowledged quite candidly: natural selection is a highly controversial concept. There are many critics (Mills and Beatty, 1979). At this point, however, I shall accept natural selection without question. In my third essay I shall return to direct examination and defend the notion against the most common criticism: that it is no genuine empirical cause but is rather empty of content, that it is, in fact, a tautology. Here, accepting selection at face value, in order to complete the scientific exposition necessary to compare evolutionary theory against physical theory with respect to the axiomatic ideal, let us note that, although selection is *the* cause of change, somewhat paradoxically at times it can act to keep the genetic situation relatively constant! Perhaps this is not really so very paradoxical after all.

Obviously, if mutation is pouring all sorts of new genes into a population and these genes put their possessors at a disadvantage to non-mutant types, then clearly what will happen is that selection will ensure that the mutants will not get a foothold, that the *status quo* will be preserved. But even in the absence of mutation, selection can keep things constant, most interestingly perhaps ensuring that a balance of different alleles is kept permanently in a population.

That selection could work in this conservative way can be seen fairly readily if we suppose that there is a selective premium on rarity (Sheppard, 1975; Maynard Smith, 1976; Dawkins, 1976). Assume that in a population we have two alleles A_1 and A_2 leading to two phenotypes P_1 and P_2, and that the chief threat to members of the population is attack by a predator (e.g., assume that we have mice and owls). Assume, however, that the predator has to learn to recognize its prey. Obviously an uncommon prey has a selective advantage over a common prey. If, therefore, allele A_1 and phenotype P_1 are rare at some initial point in time, selection favours them at the expense of A_2 and P_2. But, of course, this means that A_1 and P_1 spread and there then comes a point when there is no longer an advantage for A_1 and P_1, over A_2 and P_2. Now selection acts to hold things in a balance, with neither side able to take advantage over the other.

Another way in which one might get selection holding different genes within a population in a steady state is through a much-discussed mechanism, so-called 'balanced superior heterozygote fitness' (Dobzhansky, 1951; Lewontin, 1974). If, given two alleles A_1 and A_2, the heterozygote A_1A_2 is fitter than either homozygote, then both alleles will remain in the population in balance. Intuitively one can see this readily, because the heterozygote by stipulation will breed better than others and thus will constantly contribute both A_1 and A_2 to future generations. But even in this brief treatment we can take matters a little further than this because, fortunately, the mechanism lends itself readily to simple formal treatment.

Suppose that we have a random breeding population (i.e., without external factors Hardy—Weinberg equilibrium would obtain), and suppose also that for every A_1A_2 that breeds, proportionately $(1-S_1)$ A_1A_1's breed and $(1-S_2)$ A_2A_2's breed, and that initially gene ratios of A_1 to A_2 are q to $(1-q)$. We can, therefore, draw up the following table.

Genotype	A_1A_1	A_1A_2	A_2A_2	Total population
Adaptive value	$1-S_1$	1	$1-S_2$	
Initial frequency	q^2	$2q(1-q)$	$(1-q)^2$	1
Frequency after the selection	$q^2(1-S_1)$	$2q(1-q)$	$(1-q)^2(1-S_2)$	$1-S_1q^2-S_2(1-q)^2$

This means that the rate of change, Δq, of the frequency of the A_1 genes in the population in one generation is:

$$\Delta q = \frac{q(1-q)\ [S_2(1-q) - S_1 q]}{1 - S_1 q^2 - S_2(1-q)^2}.$$

Suppose now that the selection has done all that it can, that is suppose that we have a point where rate of gene-ratio change is zero: $\Delta q = 0$. Then $q = S_2/(S_1 + S_2)$. What this means in effect is that the alleles will balance each other indefinitely in the population, with neither allele being excluded by the other. (It can be seen that were either homozygote to become as fit as the heterozygote, then the allele of the other homozygote would vanish.)

Interestingly, one of the best-documented cases of balanced heterozygote fitness in action occurs in our own species, *Homo sapiens* (Raper, 1960; Livingstone, 1967, 1971). In parts of West Africa, native tribes are much affected by malaria. However, some people do have a natural genetic immunity to the disease, this immunity being caused by heterozygous possession of a certain haemoglobin-producing allele. One has at least a third better chance of surviving with the allele that without (i.e., $S_1 = \frac{1}{4}$). Unfortunately, homozygotes for the allele die in childhood from anaemia (i.e., $S_2 = 1$). Without reflection one might think that so severe a genetic handicap would be eliminated fairly rapidly from populations by selection. But to the contrary, so-called 'sickle-cell' anaemics keep recurring in each generation (Figure 1.2). And it is fairly clear that the reason is that the superior fitness of the heterozygotes balances out the inferior fitness of the homozygotes. Indeed, one can calculate what percentage in each generation would be expected to suffer from anaemia.

$$q = \frac{1}{\frac{1}{4} + 1} = \frac{4}{5}.$$

Hence, in each generation, $\frac{4}{5}$ of the alleles will be H (i.e., normal alleles) and $\frac{1}{5}$ S (i.e., the so-called 'sickling' allele). Thus, by the Hardy—Weinberg law, proportionately one expects $(\frac{4}{5})^2$ HH homozygotes, $2\frac{4}{5}(1-\frac{4}{5})$ HS heterozygotes, and $(1-\frac{4}{5})^2$ SS homozygotes. And in fact, one does find the expected 4–5% in each generation who are doomed to die from genetically caused anaemia.

Returning now to our own question, namely whether evolutionary theory mirrors physico-chemical theory in being hypothetico-deductive, the answer follows readily and positively. The science we have just been examining fits the pattern exactly. We have general statements, put forward as universally necessary claims (i.e., laws), which are then bound together in a deductive

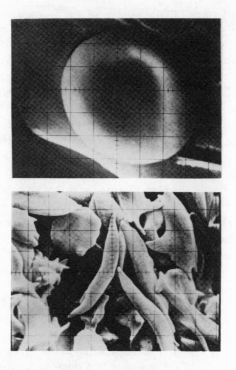

Fig. 1.2   Comparison between normal red blood cell (above, at 10 000 × magnification) and sickled red blood cell (below, at 5000 × magnification). From G. W. Burns, *The Science of Genetics* (New York: Macmillan), p. 33.

system. The Hardy—Weinberg law follows from Mendel's first law (together with a number of elementary mathematical propositions), and then claims like those just discussed above, about balanced heterozygote fitness, follow from the Hardy—Weinberg law taken together with other subsidiary assumptions. Furthermore, in the upper-level statements we get reference to theoretical entities, the genes, which are then being used to explain the deductively derived lower-level statements, which are about observable phenomena, the phenotypes of organisms. The sickling gene S is a theoretical entity, and this is used to explain the very-visible child dying from anaemia, the observable phenotype. And expectedly one finds bridge principles or translation rules. For instance, in the above mentioned discussion of blood-groups, we are told quite explicitly which alleles give rise to which groups and how this all relates to experience.

According to the theory there are two allelomorphic [allelic] genes M and N, either of which determines the presence of the equivalent antigen on the red cells.

Genotype	Phenotype or Group
MM	M
MN	MN
NN	N

(Race and Sanger, 1954, 54)

Earlier, the authors describe agglutination tests for the detection of antigens: "A serum containing a known antibody is added to a saline suspension of red cells. If the cells carry the equivalent antigen they are agglutinated; if no agglutination occurs it is concluded that the cells lack the antigen" (Race and Sanger, 1954, p. 3). All in all, I suggest we have a paradigmatic case of a theory like a theory of the physical sciences. There is nothing at all exceptional here.

1.4. BUT IS GENETICS REALLY PART OF EVOLUTIONARY THEORY?

At this point, the critical reader whose patience I begged earlier, will be able to contain him/herself no longer. There will be an outburst which will probably run somewhat along the following lines: "This is all very well what you have shown. The theory you have been expounding is indeed very much like a theory in the physical sciences. We grant, therefore, that inasmuch as your aim is to show that there is some theory in biology which is like a physical theory, you have succeeded, although it is perhaps worth noting that you have demonstrated a statistical theory like thermodynamics, rather than a non-statistical theory like Newtonian astronomy (Hull, 1974; Ruse, 1977). But, no matter. Thermodynamics is a genuine theory. It is perhaps also worth noting that your presentation hardly inspires the belief that there is anything very rigorous about the deductive nature of biology. The arguments were almost insouciantly informal. But again, no matter. Physics is often not very formal, and no doubt deductive rigour could be introduced if required. However, there is a major objection to your all-too-confident conclusions, namely that you have not proven what you set out to prove! You said that you would prove that modern evolutionary theory, the synthetic theory, is hypothetico-deductive and so forth. But you have shown this for another area of science, namely so-called 'population genetics'. As mentioned earlier, evolutionary theory is about the sorts of things you find in the evolution displays in museums, that is, about fossils and dinosaurs and so forth, and about events taking place over millions and millions of years. But you have been talking about organisms alive today and about things and changes which occur over one or a few generations. This is not evolution at all. You are a bit of a fraud."

Perhaps so. However, whatever my personal morality, I suspect that philosophically and scientifically speaking the case is not quite as black as my not-so-very-imaginary critic supposes. (See Goudge, 1961 for articulation of the kind of objection sketched above.) Let me make two, interlocking points. First, the critic rather confuses the *theory* of evolution with the *fact* of evolution (Ruse, 1973a). The fact of evolution is that for the past three-and-a-half billion years or so organisms have been evolving up from the most minute and simple forms to the complexity and diversity we see today (Ayala and Valentine, 1979). This has all taken much time, and obviously one must refer to the fossil record to infer the particular paths taken, the 'phylogenies'. But a phylogeny is not a causal theory (Carter, 1951). The latter is something which tells us not what took place, but *how* it took place. And in such a theory we want mechanisms, forces, causes, explanations; not just descriptions of particular events. This providing of mechanisms and causes is what the theory of evolution attempts to do. In a sense therefore, the theory is ahistorical, not tied to particular phenomena and events. The important point is that, if we are looking at evolutionary processes, and this is what the synthetic theory does look at, then it is mechanisms we want, not just 'evolution' *per se*. The demand for talk about the fate of the dinosaurs is misconceived. The theory might indeed be used to help to explain what happened to the dinosaurs, but it is not its job to give a simple description of dinosaur life and death.

But this all leads on to the second point. I admit readily that as yet I have not presented the whole of the synthetic theory. There is much more to the synthetic theory than what we have seen so far — its reach extends to systematics, embryology, behaviour, paleontology, and so forth. The crucial point, however, is that inasmuch as the theory moves into any one of these areas, it does so from a background of population genetics. What I am saying, therefore, is that the science I have been presenting, population genetics, which in turn comes out of the Mendelian genetics of the individual, is not separate from evolutionary theory: it is part of it. But more than this. It is the most essential part: it acts as a unifying *core* which can then be used to illuminate all other areas of evolutionary biology.

How can this be so? Simply because it is the key assumption of today's evolutionist that the largest of evolutionary changes ('macro-changes'), say from fish to humans, can be analysed ultimately in terms of the smallest of evolutionary changes ('micro-changes'), the subject of population biology. Theodosius Dobzhansky, more than anyone responsible for the development of the synthetic theory, states this fact quite explicitly.

Evolution is a change in the genetic composition of populations. The study of mechanisms of evolution falls within the province of population genetics. Of course, changes observed in populations may be of different orders of magnitude ranging from those induced in a herd of domestic animals by the introduction of a new sire to phylogenetic changes leading to the origin of new classes of organisms. The former are obviously trifling in scale compared with the latter. Experience shows, however, that there is no way toward understanding of the mechanisms of macro-evolutionary changes, which require time on geological scales, other than through understanding of micro-evolutionary processes observable within the span of a human life-time, often controlled by man's will, and sometime reproducible in laboratory experiments. (Dobzhansky, 1951, p. 16.)

1.5. THE CONSILIENT NATURE OF EVOLUTIONARY THEORY

The reader will quite possibly have anticipated that because I (like evolutionists!) see population genetics functioning in this way, the case follows readily for the third point of similarity that I hypothesized between the synthetic theory and the best theories of physics and chemistry. The theory is consilient, in that many different areas of biological science are brought together and subsumed beneath a number of powerful unifying premises, namely those of population biology. Diagrammatically the situation can be represented as shown in Figure 1.3. I might add in passing that I doubt the consilient nature of evolutionary theory is all that fortuitous. Although Darwin was ignorant of the true principles of heredity, in the *Origin* he combined his own speculations with his key mechanism of natural selection in order to make a central unifying explanatory core, which he then applied to a number of subsidary areas — paleontology, biogeography, embryology, and so forth (Ruse, 1975a). Moreover, he was quite conscious of — not to say pleased about — the consilient nature of his theory (Ruse, 1979a). In this respect, Darwin was a true follower of William Whewell, with whom Darwin had had much intellectual contact when he was a young man (Laudan, 1971; Ruse, 1975b). I suspect after the *Origin* however, neither man was too keen to advertise the earlier intimacy. Reputedly, Whewell refused to allow a copy of the *Origin* on the shelves of the library at Trinity College, of which he was Master!

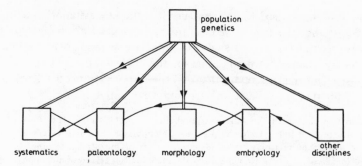

Fig. 1.3. In this figure, the rectangles represent various disciplines; the double lines the links between population genetics and other areas — such links are actually supposed to exist; and the single lines links between the subsidary disciplines — although such links do exist, those shown in the figure are just illustrative, they do not necessarily denote particular instances (from Ruse, 1973a).

Returning to the modern theory of evolution, what I am suggesting, therefore, is that workers in all fields of evolutionary studies use population genetics as background theory, as they try to formulate theories and explanations for the different problems they face. Of course, particular fields have hypotheses and assumptions peculiar to themselves as well and, in any one field, one is quite likely to find that some of the premises were first derived as conclusions in other fields; but the unifying link is the gene and its dynamics in populations. Because of the magnitude of the problems, because so much information is hard or impossible to attain, because evolutionary studies are in respects still in their infancy, one has to concede that the links are often very loose and speculative. However, for the moment, reverting to and completing my second claim about the structure of the whole theory, I submit that it is proper to think of the hypothetico-deductive model as the ideal. Foundational premises are drawn from population genetics, itself axiomatic, and then inasmuch as possible it is shown how particular premises in particular fields have to follow.

Thus, if we turn to work in paleontology what we find is that although the paleontologist is working with problems of evolution involving large events spread over great periods of time, when it comes to understanding it is the modern processes of gene transmission and spread which are crucial (Simpson, 1944, 1953; Dobzhansky *et al.*, 1977). Inasmuch as we can understand the evolution and extinction of the dinosaurs, we have to consider them as no less subject to the principles of population genetics than are animals and plants today. We shall see specific evidence of this in the next essay when we discuss some recent work in paleontology. Similarly, if one looks at something like animal behaviour and its scientific study, we find that workers in the field rely on principles of population genetics and its extensions as they try to solve their problems: Why do we get sterile workers in the Hymenoptera (ants, bees, and wasps), and why are the workers always female and never male? (Hamilton, 1964a, b; Wilson, 1975a; Oster and Wilson, 1978) Why do male fish frequently have full responsibility for the young? (Dawkins, 1976) Why do human females tend to give birth to more females than males during times of famine and hardship? (Trivers and Willard, 1973) Questions like these will occupy us in later essays and there we shall see fully the importance of the principles of genetics for students of animal behaviour, the so-called 'sociobiologists'.

One could go on listing areas in the evolutionary family, noting their reliance on population genetics. But rather than simply giving a list of flat assertions or promissory notes, by way of illustration of my point about the importance of population genetics, let me take one detailed example from one area of evolutionary studies, namely the problem which is posed to bio-geographers by the nature and distribution of organisms on oceanic islands. Organisms here on earth generally do not present gradual variation from one form to another. Rather, they are divided into more or less discrete units, 'species', where members can breed between themselves but not with those outside the group (Mayr, 1963; 1969a). Humans breed with humans, and not with apes, or birds, or vegetables. On oceanic islands what one often finds are species of organisms phenotypically quite similar to those organisms on the nearby mainland, but distinct nevertheless. And when one has groups of islands, often one finds similar but distinct species from island to island (Ayala and Valentine, 1979). Moreover, the types of island inhabitants are worthy of note. Rarely, if ever, does one find mammals, other than those that humans have introduced. Island dwellers tend to be amphibians and insects and birds. Additionally, the forms of the actual inhabitants are interesting: for instance, insects on islands are often similar to mainland forms but wingless or with limited powers of flight (Dobzhansky, 1951).

The most famous case of such island inhabitants is that which pushed Charles Darwin towards evolutionism, namely the tortoises and finches of the Galapagos Archipelago, a group of Pacific islands on the equator, some distance from the South American mainland (Darwin, 1859; Ruse, 1979a). When Darwin visited the islands, as part of his five-year, Earth-circling voyage aboard HMS *Beagle*, he did not realize at first that the tortoises and birds on different islands, although similar, tend to be specifically distinct (Figure 1.4). Later

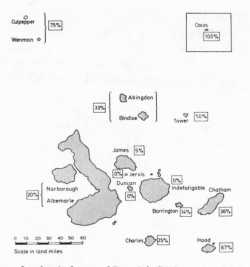

Fig. 1.4. Percentage of endemic forms of Darwin's finches on each island showing effect of isolation (from Lack, 1947).

Darwin acknowledged that the Galapagos tortoises and finches do comprise many different species and he reasoned that such an odd distribution cannot be due to mere chance; thus Darwin was led to hypothesize that a number of ancestors must have come from the mainland and evolved into the forms we find today: "We can clearly see why all the inhabitants of an archipelago, though specifically distinct on the several islets, should be closely related to each other, and likewise be related, but less closely, to those of the nearest continent or other source whence immigrants were probably derived" (Darwin, 1859, p. 409).

Darwin's explanation has stood the test of time, for his is essentially the position accepted by evolutionists today (Lack, 1947; Mayr, 1963). It is argued (generally for oceanic islands and specifically for the Galapagos) that we get a few founders from the mainland. Because of the difficulties of travel, this is the reason why these founders tend to be birds and insects and marine organisms and the like, and not organisms like mammals which obviously cannot fly or swim such distances. Then, because of new strange conditions one gets evolution to new forms — forms which are often specially adapted to island life. Insects without wings are less liable to be blown out to sea and lost than insects with wings. And then, occasionally, one gets passage from one island to another, and the whole evolutionary process starts again: the distinctive nature of particular island conditions often influencing the actual form of evolution. In the case of Darwin's finches note how Charles, Chatham, and Hood have a higher proportion of endemic forms than one might have expected from their geography. This is because the trade-winds blow in such a way as to make them relatively inaccessible to birds from the central islands (Lack, 1947).

Now, the point is that in all of this, explaining the presuppositions of population genetics are absolutely crucial. When an organism arrives on an island, or rather when a group of organisms arrive on an island (this could be one pregnant female), and breeding starts, the assumption is that there will be all kinds of new selective pressures pulling these organisms away from their parent populations. But if regular genetic principles do not hold, then selection could have no effect, nor would one expect the appearance of adaptive characteristics suitable for the newcomers in their island habitats. What biogeographers believe is that the Hardy—Weinberg law was just as valid for the first finches to arrive on the Galapagos, as it is for any population around us today, including ourselves! Because of the background stability provided by the law, selection (and also mutation) can take effect and thus the evolutionary process can proceed.

Biogeographers believe, in fact, that population genetics can paint in even more of the picture than this. We have seen that it is believed that different alleles can be held in a population by selection – through the advantage of rareness, through superior heterozygote fitness, and so forth (Dobzhansky, 1951; Lewontin, 1974). This means that, in the parental mainland population, there is no such thing as an average member, because all members will probably have some alleles that others do not have, and not have some alleles that others do have. This fact, something drawn from our knowledge of population genetics, has been the basis of a brilliant hypothesis formulated by Ernst Mayr: the founder principle (Mayr, 1942. 1963). Mayr points out that the founders of a new population will necessarily be atypical, because of the variation in the parental population and, therefore, when they reach their new home, there will be a fairly rapid genetic 'shakedown', as the limited number of alleles in the founder group try to form new viable genotypes. Genes do not exist in splendid isolation – they affect each other and they mutually affect phenotypes. But this means that "no gene frequency can be changed, nor any gene added to the gene pool, without an effect on the genotype as a whole, and thus indirectly on the selective value of other genes" (Mayr, 1963, p. 269).

Put matters this way. Suppose in a parental population, one homozygote

A_1A_1 was generally far less fit than the homozygote A_2A_2, but survived because of a certain rarity value. In the founder population, there may be no A_2 alleles at all! Thus, all the members are A_1A_1 homozygotes. This could affect alleles at other loci and, at the phenotypic level, lead to all kinds of strange selective forces, which would not trouble the members of a larger population with more genetic variety to draw on. The result of all this could be that initial evolution would be rapid and lead to fairly significant differences in the new group from the old group. And although, in a sense, there is a randomness about all of this, due to the way in which atypical species members form the founding parents of a new group, ultimately it all comes back to selection and genetics, which latter show that the founding parents are bound to be atypical because all species members are atypical!

I shall not pursue the example further. The point is made. Population genetics is the base on which the whole biogeographical explanation of the distinctive nature of organisms to be found on oceanic islands. Students in the area rely absolutely on such principles as the Hardy—Weinberg law and those which show how such forces as selection and mutation can affect gene ratios. And this reliance is found repeatedly through the evolutionary spectrum. Population genetics unifies the different areas. In this respect, evolutionary theory is as consilient as any theory in the physical sciences. (For a somewhat different perspective on evolutionary theory, see Beckner, (1959) but see also Schaffner (1980).)

1.6. CONCLUSION

I argue, therefore, in conclusion that in three essential respects we have seen significant similarities between the modern synthetic theory of evolution, neo-Darwinism, and the theories of physics and chemistry. I admit to much looseness and many gaps in evolutionary work, but the same basic ideals are there and manifested. As noted, I am not arguing that there are no real differences between biological thinking and physico-chemical thought. I think there are and I shall look at them in a later essay. But I do argue that the time has come to stop looking on evolutionary theory as some strange inadequate freak and, at least, to admit it as a candidate to the school of genuine and important scientific theories.

NOTE

[1] I do want to emphasize that I am aware of disagreements between evolutionists. (See, for example, Gould and Lewontin 1979; although, see also Cain, 1979.) Because I am writing a philosophical essay rather than simply giving a journalistic account of the present state of evolutionary studies, I ignore disagreements here, in order to get on with my main task. But I do not think my quick dismissal affects the basic points I want to make. Objections at this stage may, however, rest on a semantic point. As will become clear in the next essay, where I am prepared to include within the synthetic theory recent new hypotheses about the rate of evolutionary change, I do not take a narrow view of modern evolutionary theory. Hence, I would like to think that what I say applies to virtually all active evolutionists.

I confess, however, that recently some evolutionists have stretched their Darwinian commitments to the limit. See, for instance, S. J. Gould, 'Is a new general theory of evolution emerging?' *Paleobiology* 6 (1980), 119–30.

THE EVIDENCE FOR EVOLUTIONARY THEORY

In this second essay I want to answer a question which one hears frequently about the synthetic theory of evolution and which, I am sure, many who have read my first essay find urgently pressing. Suppose we grant everything I have, so far, claimed about evolutionary theory — no doubt a major 'suppose' in some people's opinion, but one which, for the sake of argument, we can grant here. There is still the overwhelming question of truth. Is the synthetic theory simply a grandiose exercise in system building, the castle in the sky to end all castles in the sky, or can it command our respect as a reasonably well-confirmed theory? Is it fact or is it fiction?

I suspect that in the minds of many there will be little doubt as to the answer. Constantly one hears that neo-Darwinism is 'only a theory', meaning by this that, if Einsteinian relativity is your idea of a real theory, then neo-Darwinism is anything but a theory. It is stated flatly that the synthetic theory is little more than self-aggrandizing speculation, because there is no real evidence to support it. Indeed, it is suggested that evidence and confirmation are, if not theoretically impossible to obtain, virtually inaccessible in fact. Who can say why the dinosaurs evolved? We cannot go our and look at them and run tests. Nor can we employ the well-trodden path of scientific-theory-confirmation, namely predicting and seeing if the predictions obtain. No one can wait a few million years to see if the elephant will grow a longer trunk or if the giraffe's neck will shrink. All is untested and untestable hypothesis (see, for instance, Macbeth, 1971; Bethell, 1976; Manser, 1965; Barker, 1969; Lee, 1969.)

Why then, one might ask, has the synthetic theory risen to the position it has? As mentioned in the last essay, sometime the suggestion is made that the success has come virtually by default. Basically, evolutionary theory has triumphed because it "has never had to deal with serious scientific opposition" (Manser, 1965, p. 18). And even when such opposition has been raised, regretfully this opposition has been downplayed "possibly not on very adequate evidence" (ibid.). Other suggestions see more sinister (or charitably, more naive) motives behind the rise of Darwinism and neo-Darwinism. Darwin, we learn, invented a new logic, the 'logic of possibility'. "Unlike conventional logic, where the compound of possibilities results not in a greater

28

possibility or probability, but in a lesser one, the logic of the *Origin* was one in which possibilities were assumed to add up to probability" (Himmelfarb, 1962, p. 334).

Whatever the surface merits of objections like these, one suspects that they are fuelled by a deep-down distrust and dislike of Darwinian and neo-Darwinian biology, its methods, and its results. Charles Darwin may lie moldering beneath the flagstones of Westminster Abbey, but even today, he and his work stir surprising emotions.[1] Nevertheless, objections to and queries about the synthetic theory are made by those who, on every count, are sympathetic to biology and its personalities. Thus, for instance, the deservedly well-known Canadian philosopher, Thomas Goudge, feels obliged to observe that: "A philosopher surveying the pros and cons of this controversy finds himself in no position to espouse one side rather than the other. The only reasonable conclusion seems to be that considerations so far advanced do not permit a settlement" (Goudge, 1961, p. 51).

Just as I argued in the last essay that criticism of the nature and structure of evolutionary theory was misguided, so in this essay I want to argue that criticism of the evidential status of evolutionary theory is misguided. Indeed, I would go further and state categorically that such critics are absolutely mistaken. Do not misunderstand me. I do not claim that all questions within evolutionary theory are definitely decided. This is no more the case than that all major questions within physico-chemical theory are definitely decided. In a later essay (Essay 5) in fact, we shall see that there is a very lively ongoing controversy about certain aspects of the evolutionary process (Lewontin, 1974; Ayala *et al.*, 1974). My point is simply the basic one that the synthetic theory of evolution, the theory presented in the last essay, is well-confirmed in its fundamental claims. If I might indeed be allowed a prediction that neither I nor my readers will be around to check out, I predict that a hundred years from now (or say 2059, the two-hundredth birthday anniversary of the *Origin*), people will still be neo-Darwinians essentially in the way that they are today.

A key part of my case has been made already. I think a lot of the criticism of the synthetic theory occurs because people simply do not know what the theory is. In particular, there is the already-mentioned confuion between the *theory* of evolution and the *course* of evolution, phylogenies. Much will never be known about phylogenies, because the information is lost, irretrievably. And due to this, many of the events of the past cannot be explained fully — we simply do not know the relevant particular circumstances. Let me openly admit that I do not know why the dinosaurs died out and that I doubt that

anyone else does either; although on this subject, some fascinating suggestions
have been made recently (Alvarez *et al*., 1980). However, the key point is that
this inability to give full explanations of particular events like the dinosaur
deaths comes about, not because neo-Darwinism is 'only a theory', but
because we do not know all the pertinent external circumstances. Similarly,
I do not know whether the elephant's trunk will grow longer or shrink but,
again, the key failure lies in ignorance about future external forces impinging
on elephants. In short, to criticize the synthetic theory on these sorts of
grounds is just not fair. However, let us not waste our time in complaining.
Let us rather turn positively to the real theory of evolution and try to show
that there is solid evidence for this theory, that it does not succeed by default,
and that there is nothing mysterious or dishonest about the standards of
proof used by evolutionists.

2.1. EVIDENCE FOR THE SYNTHETIC THEORY'S CORE

We begin, obviously, with what I have argued is the unifying core of evolutionary theory, namely population genetics. What is the status of this? I shall
not dwell at length on three facts that seem to me to be as important as they
are incontrovertably true. The first of these is that the basic principles of
genetics at the individual level − those which tell us about the nature of the
gene, the way it is transmitted from generation to generation, the way in
which it can produce the parts needed to make the body, even the way in
which it changes − are known in great detail, and are known to be true beyond reasonable doubt. Not all queries are answered, of course, but brief
perusal of any elementary genetics text shows the solid status of the theory
of heredity. (See, for instance, Strickberger, 1975.) Given what we know,
namely that population genetics, the background for the rest of evolutionary
studies, relies in turn on individual genetics for a base, this well-confirmed
status of central knowledge about the gene is of obvious importance to the
whole synthetic theory (Li, 1955; Dobzhansky, 1970).

The second fact is one of which Charles Darwin himself made much: there
is no doubt that selection in one form or another can have great effects on
organisms. In the *Origin* Darwin discussed at great length the way in which
animal and plant breeders can mold their subjects to virtually any prescribed
form they wish. "Lord Somerville, speaking of what breeders have done for
sheep, says: 'It would seem as if they had chalked out upon a wall a form
perfect in itself, and then had given it existence'" (Darwin, 1859, p. 31).
Darwin argued that this power of artifical selection was strong analogical

evidence for the power of natural selection, and in this he was surely right (Ruse, 1973b, 1975a). If artificial selection were totally ineffectual, then natural selection would be in jeopardy; but conversely, given the great power of artificial selection, then properly one can regard more favourably any mechanism which works on essentially the same lines, namely those positing the restriction of breeding in any population to but a select few, or a least, those positing a differential reproduction (Maynard Smith, 1975). One can say this whilst admitting that the analogy from artificial selection is limited. Unless one can give reasons to suggest that the breeders' are mimicking natural forces in their work, one can draw little by way of conclusion about exactly how natural selection would work. For instance, breeders are not really that interested in producing new species, and expectedly even Darwin's friend (like T. H. Huxley) pointed out that, since artificial selection had never produced so fundamental a natural phenomenon as a species, the analogy was more suggestive than definitive (Ruse, 1979a). Today, we are further along with attempts artificially to create species barriers, but this is not to deny the limited value of the analogy (Thoday and Gibson, 1962). These limits notwithstanding, however, the analogy is an important positive contribution to the overall evidential picture.

Third, there is the indisputable fact that the physical preconditions for the operation of a mechanism like natural selection do exist. A differential reproduction means that some organisms are contributing more to the next generation than are others. But why should this differential occur at all? The answer is that equal contributions by everyone are simply not possible because, as Darwin noted, there are far more organisms being born than can possibly survive and reproduce. Hence, there must be some factors limiting potential population growth. That is, there must be some sort of differential mortality and reproduction. A modern evolutionist puts matters this way:

In a few species, such as the herring, the maximum potential increase per generation may be as much as a millionfold. However, even in species such as our own where relatively few offspring can be produced by a single pair, the potential rate of increase is very rapid. If we assume, for example, that the average number of children, born to a married couple, who themselves grow up and marry is, in optimal conditions, only four, the population will doubt in each generation. The population would then increase a thousandfold in ten generations, and a millionfold in twenty generations, or about 600 years.

Since animal and plant numbers do not in fact increase indefinitely in this manner, it follows that either not all individuals born survive to sexual maturity, or that some sexually mature individuals do not breed, or that breeding individuals produce fewer offspring than they would under optimal conditions. (Maynard Smith, 1975, pp. 32–3.)

The terrific importance of these various facts for the general theory of evolutionary processes, population genetics in particular, just cannot be denied. If one is going to reject the synthetic theory, then one had better find some other way of accounting for them. Of course, this is not to imply that accepting their importance at once validates everything that a modern evolutionist would want to claim about the workings of selection. The differential reproduction (on the facts just given) might not be systematic in the way that we saw supposed in the last essay. But the point is that certain basic indisputable facts in the organic world give one a firm basis on which to build a theory like population genetics which crucially incorporated a causal mechanism like natural selection.

Suppose you do agree with me on this point. Suppose that facts like these just given provide a background for a sympathetic hearing for a theory like modern population genetics. What more might you want before you agree that here we are looking at a reasonably well-confirmed body of knowledge? A not-unreasonable request would be for evidence of the postulated processes of evolutionary change actually at work. Once again let us take physics as a guide. Led by such a guide, what we hope to find is that two sources will prove fruitful in the search for results: human-controlled experiments, often within the laboratory, and direct observation of on-going natural processes in the wild (Hempel, 1966). In the case of the core of the synthetic theory evolutionists have gone to both sources, and they have come away with hands full.

A laboratory experiment on postulated evolutionary processes, which has become something of a classic in its own time, was one reported in 1957 by Dobzhansky and his associate O. Parlovsky. It was performed on captive populations of Drosophila in order to test the founder effect hypothesis (Mayr, 1942, 1963). You will remember that this hypothesis suggested that, because of the heterozygosity in natural populations, founders of new populations, being necessarily atypical, may well show rapid initial change as their limited range of genes coalesce into stable, cohesive genotypes. Thus, although a certain allele, say A_1, may exist quite stably at a certain ratio in a large population, in a small population the gene could be much fitter or less fit than hitherto. If, for instance, this allele A_1 exists only because it has a rarity value against a generally fitter allele, A_2, but the new population has no A_2's at all, then the allele A_1 could spread far more rapidly than previously. And this, in turn, could have a domino effect through the genotype. To test this hypothesis, Dobzhansky and Parlovsky monitored the fate of a certain gene over a number of generations in laboratory populations of *Drosphilia pseudoobscura*.

Initially they began with ten populations of flies numbering 5000 individuals each ('large') and ten populations of flies numbering 20 individuals each ('small'). Each population had a 0.5 frequency of the genetic constitution PP, and the test was run for 18 months over about 18 generations. The prediction was that, if the founder effect is true, even though at the end on average PP might be the same in large and small populations, one should get a far greater range of variation in the small populations. In some populations the genetic background will favour PP and in others it will not. And as the accompanying Figure 2.1 clearly shows, the prediction was triumphantly vindicated. I would suggest that an experiment like this has the same logical status and importance for the synthetic theory of Young's double slit experiment has for the wave theory of light.

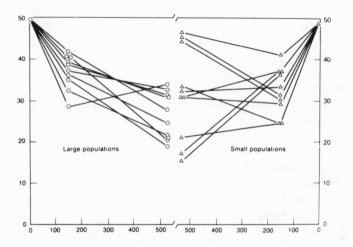

Fig. 2.1. "The founder effect in laboratory populations of Drosophila pseudoobscura. The graphs show the changes in the frequency of a certain genetic variant known as PP. Note that time proceeds from left to right for the large populations, but from right to left for the small populations, so as to facilitate comparison of the final results" (from Ayala and Valentine, 1979, p. 119; after Dobzhansky and Parlovsky, 1957).

In addition to experiments on general claims like the founder effect, evolutionists have also looked experimentally at more particular suggestions. Remember how it was hypothesized that the reason why so many species of insects on islands are wingless is because there is a selective advantage on islands in having what is normally obviously a severe handicap: insects who fly about tend to get blown out to sea, and that is the end of them!

Dobzhansky tested this hypothesis experimentally (Dobzhansky, 1951). Populations of *Drosphila melanogaster* were kept in cages, and the initial members were a mix of normal flies with wings ('wild type') and mutant flightless flies with stubby wings ('vestigal'). As expected, under normal circumstances, wild type flies proved much superior to vestigal flies, and rapidly out-reproduced them. However, when circumstances were manipulated so as to simulate island conditions – fans kept currents of air blowing through the cages – as also predicted it was the vestigal flies which proved fitter and which reproduced at the wild type's expense. Again, therefore, we have the orthodox pattern of prediction and confirmation.

Turning to the other side of the evidential coin, we find that much successful effort has gone into the observation of selective forces at work in natural populations. Once again, highlighting a classic piece of work, this time by the English, who continue to work brilliantly in the naturalist tradition of Darwin himself, we have the observations by H. B. D. Kettlewell and others on the peppered moth, *Biston betularia* (Kettlewell, 1973; Ford, 1971). In the middle of the last century the moth was always a light, speckled-gray colour. Then, particularly in industrial areas, it was noticed that the moth started to appear in a dark melanic form until, indeed, the dark form was the norm and the lighter form the rarity. The explanation put forward was that colour has an adaptive significance as camouflage. As the trees in the moth's habitat got progressively darker from pollution, it grew more and more in the moth's interests to be dark also and, thus, inconspicuous to its greatest threat, predators (Figure 2.2). Kettlewell confirmed this fact, at the same time showing that in unpolluted areas of Britain, the moth still exists in the light form and, even more importantly, showing that the grave threat to the moths is indeed predators, namely birds. Through observation he was able to establish that light forms tend to get picked off and eaten against dark trees, and dark forms tend to get picked off and eaten against light trees. Combining observation with experiment, Kettlewell reinforced his finding by releasing marked light and dark moths in areas of high and low pollution. Expectedly, far more dark moths could be recaptured in polluted areas and far more light moths in unpolluted areas. (See Table 2.1) Rounding out this story let me note that the peppered moth is not unique: industrial melanism has been discovered in over a hundred species of moth (Ayala and Valentine, 1979). It seems, therefore, not to be some unique phenomenon, which could be due to change or to factors unrelated to predation and camouflage. Also, there is a happy ending to the story – for light-coloured moths at least! Where strict pollution controls have been enacted, the trees get cleaner and expectedly the light

(a)

(b)

Fig. 2.2 (a) The peppered moth *Biston betularia* and melanic form against a soot-covered tree. (b) The same forms against a clear, lichen-covered tree. (Courtesy, the estate of H. B. D. Kettlewell.)

TABLE 2.1
Biston betularia moths recaptured in two areas

Locality	Moths released		Moths recaptured	
	light	dark	light	dark
Birmingham	64	154	16 (25%)	82 (53%)
Dorset	393	406	54 (13.7%)	19 (4.6%)

Birmingham is badly polluted and Dorset has virtually no pollution. (From Ayala and Valentine, 1979, p. 133; after H. B. D. Kettlewell.)

moths reappear in increasing numbers. The selective pendulum has started to swing back again.

Possibly at this point it will be objected that the change of the magnitude involved in *Biston betularia* is nowhere like great enough to prove evolutionary theory, that is, gradual evolution through natural selection (Bethell, 1976; Macbeth, 1971). What we need is direct evidence of apes turning into men, or the like. I doubt, however, that this objection is fair. At the moment we are not looking at the whole theory, but rather at the central core: population genetics. Given that this core does indeed deal in small changes, the case of the melanic moths is highly relevant. If one had no such cases, then the core would be thrown into question. Of course, this is not to concede to the critics that the moths have no relevance for the overall theory. Failure of the core would mean that indirect queries would be directed towards the whole synthetic theory. Conversely, however, as examples like that of the moths boost the core, indirectly they, thus, support the whole theory.

But still the persistent critic might ask for more. Continuing to restrict ourselves to the core of the synthetic theory, even if we grant by the very nature of the case that we are going to observe only limited change, surely we might hope to see some of the qualitative changes postulated by the core of evolutionary theory, as well as the quantitative changes? In particular, the theory postulates that new groups of reproductively isolated intra-breeding populations, species, come about through processes postulated and backed by population genetics. What about some observational evidence for this? There seems to be nothing in the synthetic theory which puts this beyond direct human sense; but until we have such evidence, the power of even the central core of evolutionary theory seems somewhat hypothetical.

Obviously, if one agrees that the founder effect is important in the formation of new species then, as we have just seen, it is not quite true to say that we have no evidence at all of a qualitative change like the formation of a new species. But rather than delaying to squabble about how much evidence one needs or has to have to establish the truth of the core of evolutionary theory, let me counter the critic directly by revealing that there is, in fact, some direct evidence of new species being formed in nature, and that, moreover, some of the factors postulated by population genetics seem to be involved. In particular, S. Prakash has put together convincing evidence that a certain population of fruitflies (*Drosophila pseudoobscura*) in Bogota, Columbia, is right in the middle of evolving into a new species (Prakash, 1972; Lewontin, 1974). In 1955 and 1956, extensive collection in Columbia revealed absolutely no individuals of *D. pseudoobscura*. By the nature of the case, one cannot say absolutely that there cannot possibly have been any of the insects in Columbia in the mid-1950s, but the chances are certainly strongly against it. The nearest population was in Guatemala, 1500 miles away. However, in 1960, *D. pseudoobscura* started to appear in traps in Bogota, and before long in some localities it was as numerous as indigenous fruitflies. What makes this insect particularly exciting and pertinent to us is the fact that already the Columbian *D. psdudoobscura* seems to be on the way to being a new species. Tests have shown that sterility barriers are in the process of being erected between Columbian *D. pseudoobscura* and members of the species collected from elsewhere — females from Bogota produce totally sterile males when crossed with *D. pseudoobscura* from other localities. The reciprocal cross produces normal sons. We are part way to new species.

If, indeed, population genetics is to throw any light on what is going on here and is, in turn, to receive support from the phenomenon, then one thing we might certainly expect is some evidence of the founder effect in action. And it is a pleasure to be able to report that there is such evidence. Study of the chromosomes suggests strongly that the closest relatives of the Bogota flies are those in Guatemala. Expectedly, the flies in Bogota as a group show far less intra-population variation than do the flies in Guatemala — this follows from the fact that the few founders would necessarily have been able to bring only a limited sample of their parents' total variation. Expectedly, at those loci, where the parents have little or no variation, or where there is one allele form much in the majority, the Bogota flies tend to go with the parents and expectedly, the Bogota flies give evidence that the founders cannot have been totally typical; because no one is totally typical. One allele in particular is almost unknown in the parent population (1—2%) but is the norm in the Columbian flies (87%). All in all, what we have here is a species in the making, with features predicted by (and thus confirming) population genetics.

I shall not labour the general point any further. I do not claim that everything in population genetics is or should be accepted as gospel truth. But at
the level at which we are working, with genes that cause appreciable effects
on phenotypes, information from all directions, direct and indirect, experimental and observation in the wild, points to a cohesive body of theory with
strong empirical backing. The synthetic theory of evolution rests on a solid
core.

2.2. EVIDENCE FOR THE WHOLE THEORY

What about the synthetic theory taken as a whole? Even if we grant that
there is evidence for the core part of the theory, is it reasonable to accept the
overall theory or should we at least suspend judgement? Unlike the critics, I
certainly do not want to argue to an extreme position, claiming, for instance,
that every part of the theory is as strong as every other part, or that every
particular explanation is as good as every other one. However, I would suggest
that when we look at the overall picture, we see a body of theory and evidence which, in the language of the courts, establishes 'beyond reasonable
doubt' the nature and the causes of the evolutionary process – natural
selection working on random mutation leading eventually to well-adapted
organisms. What we have throughout the range of evolutionary disciplines is
the same dynamic explanatory/confirmatory interplay that we find in the
physico-chemical sciences. On the one hand, the theory can and does throw
explanatory and predictive light on the facts of the organic world, whether
they be general or particular and, on the other hand, these explained and
predicted facts increase our confidence in the essential truth of the theory,
both in its core and in the subsidiary assumptions that occur in any particular sub-branch, like biogeography and paleontology.

As in the last essay, space limitations force me to be selective and, in any
case, I have neither reason nor desire to duplicate the many excellent texts
on evolutionary theory. (See, particularly, Mayr (1963), Maynard Smith
(1975), Dobzhansky et al. (1976), and Ayala and Valentine, (1980).) I have
mentioned already that one area where today some of the most exciting
advances in evolutionary understanding are occurring is in the science of
animal social behaviour, 'sociobiology'. As promised, in later essays I shall
be looking at some of the proffered explanations, the evidence for which will,
I shall suggest, add to our general confidence in the overall theory of evolution. (To forestall premature criticism, I am referring here particularly to
work in the animal world, and not simply to the notorious and controversial

sociobiological excursions in the human sphere.) Another area where one might turn in order to illustrate the general point is paleontology, where also today we see exciting new ideas at work. Later in this essay, indeed, I shall be giving instances of paleontological work which, I believe, do support the general thesis I am trying to establish. Here, however, let me short-cut exposition by returning to the example of the last essay, namely the science of biogeography in general and the problem of oceanic island speciation in particular.

Clearly, if what was argued in the last essay is well-taken, namely that the synthetic theory throws valuable explanatory light on a phenomenon like the nature and distribution of Darwin's finches on the Galapagos Archipelago, then conversely we have confirmatory evidence for biogeography in particular and, for the overall synthetic theory, in general. Take the most striking fact about the distribution of species, namely that the central islands of Duncan and Indefatigable have no endemic forms at all, but that as one moves outwards one gets more and more such forms, until one gets to Culpepper and Wenman which have 75% endemic forms and Cocos Isle which has but one species, found nowhere else (refer back to Figure 1.4). If this is not strong evidence for the process of gradual evolutionary change, as postulated by the synthetic theory, I do not know what is. The theory predicts that one would get such a spreading out of species, and this is what we find in confirmation.

But there is more to the story than this. Although the founder effect, with the element of randomness it introduces, is believed crucially important, the key distinguishing factor in natural selection is that it does not produce just any kind of organic features. Rather, organisms through their various characteristics are going to be adapted to their circumstances. Organic features are going to help in the struggle to survive and reproduce. Thus, in looking at Darwin's finches, or at any comparable group like the Hawaiian honeycreepers (Drepanididae), or the many species of *Drosophila* endemic to the Hawaiian islands (about one third of the world's 1250 described species are endemic to the islands), what the synthetic theory leads one to expect is that species' differences will reflect different adaptations. Different Darwin's finches, for instance, will be adapted for success in different fields or areas – and where species exist together on the same island, differences will be accentuated because the interspecific rivalry will sharpen contrasts, wiping out intermediates without extreme specializations and, perhaps, driving different populations in different directions, thus avoiding competition (Dobzhansky *et al.*, 1977).

This is precisely what we find, as predicted, thus as explained and, hence, as confirming. Darwin's finches are a classic case of 'adaptive radiation', where essentially similar birds have diversified to take advantage of different ecological niches, different foods, and so forth. Some birds are cactus-eating, some birds are insect-eating, some birds are omnivorous. And their features give testimony to this fact. Thus, considering the beaks (Figure 2.3), we see

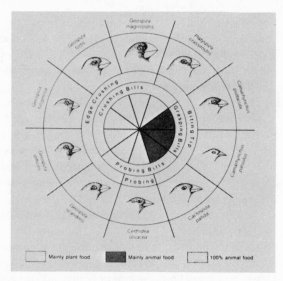

Fig. 2.3. Darwin's finches. These ten species are all found on Indefatigable Island. As can be seen clearly, different species have different beaks adapted to their different foods (from Dobzhansky *et al.*, 1977, p. 187).

that some beaks are stubby and strong, just what one needs to crush up tough vegetable food. Others have more delicate beaks which are good for digging or probing into things, like the bark on trees, just what one needs if one is searching for insects. Expectedly also, even within the species, we find differences reflecting the need for and pressure towards adaptive excellence. If on one island, a particular kind of finch is particularly well adapted to a certain mode of feeding, finches of other species tend not to compete. But if such a kind is absent on other islands, finches of those other species tend to evolve in such a way as to take the absent kind's place. Thus, speaking of the insectivorous tree-finches *Camarhynchus psittacula* and *C. parvulus*, and the woodpecker-finch *C. pallidus*, the ornithologist David Lack observed:

On James and Indefatigable all three species occur together. But on Chatham the large insectivorous tree-finch *C. psittacula* is absent, and here . . . the small species *C. parvulus* attains an unusually large size. But this is not all, for the Chatham form of the wood-pecker-finch *C. pallidus* has a shorter beak than usual. This suggests that, in the absence of *C. psittacula*, there has been survival value to *C. pallidus* in becoming less specialized in beak, and that the Chatham form of *C. pallidus* takes some of the foods which on other islands are taken by *C. psittacula*.

The opposite situation perhaps occurs on Abingdon and Bindloe, where the wood-pecker-finch *C. pallidus* is absent, and the large insectivorous tree-finch *C. psittacula* has a longer and straighter beak than usual, suggesting that it may take some of the foods normally taken by *C. pallidus*. (Lack, 1947, pp. 70–1.)

I cannot overemphasize that these are precisely the kinds of results that one would expect were natural selection at work. Conversely therefore, I argue (with the evolutionists) that results like these provide strong and satisfying evidence for the active force of natural selection. When one takes these kinds of facts about Darwin's finches — made that much more striking by closely analogous phenomena like Hawaiian honeycreepers and *Drosophila* — and combines with them the already-discussed evidence for the core of the synthetic theory, population genetics, it really does not seem reasonable to conclude that the basic picture is open to reasonable doubt. I am *not* saying that no important questions remain. In respects, we are only starting to scratch at some of the problems to do with evolutionary processes, just as we can say that, in a sense, many of the great theories of physics and chemistry in their early years only started to scratch at many of their problems. But to deny the explanatory relevance of natural selection to the beak-differences between (say) the insectivorous *Certhidea olivacea* and the vegetarian *Geospiza magnirostris* or, conversely, to deny the confirmatory relevance of these beak-differences to natural selection (and the theory in which it is embedded), is surely to take open-mindedness beyond the point of common-sense.

I hope that I have managed to carry the reader this far with me. I certainly hope that I have managed to carry the reader whose mind was not irreversibly closed to evolutionary theory before this essay began. However, even the sympathetic reader may by now be feeling that by now the time has come to give the opposition, if not equal space, at least some opportunity to air their views. And this feeling becomes the more urgent because, as we have seen, it has been objected that the very reason for the synthetic theory's apparent success is the simple fact that its supporters unfairly refuse to take seriously rival theories of organic origins. Therefore, out of fairness, not to mention out of respect for sound scientific methodology, we ought to assess the strengths or rival attempts to explain organic existence and diversity, both as we see it around us today and as it is revealed to us by the path that it has left behind it. Let us, therefore, turn next to the question of alternatives to the synthetic theory. I argue that there have indeed been such alternatives, that they have been examined fairly on their scientific merits, and that they have been dismissed − rightly. In the struggle for existence between rival evolutionary theories, the synthetic theory proved itself the fittest, on its own merits.

2.3. RIVALS: THE FIRST CHAPTER OF GENESIS

I suppose really that, if we broaden our net wide enough to consider rival theories of organic origins as opposed simply to rival theories of evolution, we ought to begin by looking at one or more variants of Special Creationism, meaning by this the claim that all organisms including humans were created instantaneously, miraculously by God. This hypothesis, if I may so call it without prejudice, could take the conservative form, where the age of the earth is fixed at around six thousand years (a figure arrived at by extrapolating from the long lists of geneologies given in the Old Testament), and where supposedly everything was created at the beginning in six literal days of twenty-four hours each. This was the position accepted before the nineteenth century and which still finds much favour in many parts of North America today (Gerlovich *et al.*, 1980). Or one might promote a more liberal Special Creationism, allowing for a greater length of time, perhaps interpreting the six days as six long ages, and supposing God to have created on an ongoing basis. This was the kind of view accepted by many scientists in the first half of the nineteenth century, before Darwin published the *Origin* (Miller, 1856; Millhauser, 1954; Ruse, 1979a).

Although I realize that there is absolutely no possibility of changing the thinking of those already committed to Special Creationism, let me say both for the record and for the benefit of those who think perhaps Special Creationists should be given the opportunity to make their case (in the schools particularly), that if one subscribes in any honest way to the principles of science − or any kind of straight reasoning for that matter − then Special Creationism is, in the light of today's knowledge, ludicrously implausible. All the evidence of geology, backed by physics and chemistry, is that the earth is many orders of magnitude greater than 6000 years − rather more than 4½ billion years is the current estimate − and today, life on earth is believed to go back well over 3 billion years (Maynard Smith, 1975; Dickerson, 1978; Schopf, 1978). For all the imperfections of the fossil record, it is quite complete enough to point overwhelmingly to an evolution from primitive forms to today's sophisticated diversity (Bowler, 1976; Valentine, 1978). And extant life bears abundant testimony to its evolutionary origins. The facts of embryology and comparative anatomy in particular are scientifically inexplicable without evolution. Why otherwise should a dog and a man have identical embryos, or why should the arm of a man, the fore-leg of the horse, the flipper of the dolphin, the arm of the mole, the wing of the bat — every one of these parts used for a fundamentally different thing — nevertheless be formed from the same set of bones adapted to their different end? (Figure 2.4). Without evolution, these isomorphisms ('homologies') simply do not make sense.

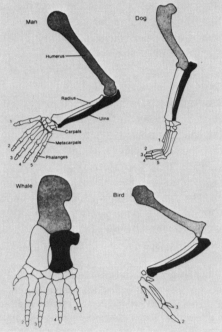

Fig. 2.4. Homology between forelimbs of different vertebrates. Numbers refer to digits (from Dobzhansky *et al.*, 1977, p. 264).

Please note that I make these claims assuming that one has made a commitment to the primacy of observation and scientific inference. Obviously, if one accepts the primacy of the literal word of the Bible, then the claims and supporting arguments of science are irrelevant. It would be like trying to convince someone to obey the rules of cricket, when they are playing baseball. But it is worth noting that many of today's Special Creationists believe that they can have scientific cake and eat it too. They believe that not only can they accept every word of the Bible, but also that they can justify their position by reference to scientific facts and inferences. At least, they write very long books to this effect! (Whitcomb and Morris, 1961; Gish, 1972; Morris, 1974). However, if one studies these works carefully − or not so carefully for that matter! − one soon discovers that the argumentation is tortuous, to say the least, which I suppose is precisely what one would expect, given the conclusions being derived. Attention is very selective: for instance, long chapters are given to supposed gaps in the fossil record, whereas subjects like geographical distribution and comparative anatomy, subjects which from Darwin onwards have been those most prized by evolutionists, are given very scant treatment. The works of eminent evolutionists have been scanned with care and isolated comments are quoted right out of context. Perusal of creationist works give the impression that the leading spokesmen for the cause are Mayr, Simpson, and Dobzhansky. There seems something incongruous to the point of dishonesty in this willingness to accept the occasional queries, doubts, and side-comments of evolutionists, and yet to reject absolutely every one of their main claims. If these men are so untrustworthy in essentials, what ground is there for referring to them at all? And worst of all, the argument of evolutionists tend to be distorted to the point of caricature. Let me quote one of the few creationist discussions of comparative anatomy that I have found included in a putative public schoolteachers' guide to the teaching of creationism. So that I shall not fall into the sin of those whom I criticize, I give the whole passage:

Similarities in Morphology (*Comparative Anatomy*)
Similarities in structure are considered one of the main evidences of evolution. To some extent, since the standard Linnaean classification scheme is arbitrary and man-made, such similarities may actually indicate common ancestry. This is certainly true at the level of varieties, and possibly also at the species level and occasionally at the level of higher categories. It should be remembered, however, that no observational or experimental evidence exists for ancestral relationships in these higher categories. This is purely an evolutionary assumption.

Probably the leading American taxonomist (*taxonomy* is the science of classification) is Ernst Mayr, of Harvard. Professor Mayr emphasizes that all such higher categories (genera, families, orders, etc.) are quite arbitrary, since no experimental proof can be offered to demonstrate any such relationships. A reviewer of Mayr's most authoritative work, *Principles of Systematic Zoology* (New York, McGraw-Hill, 1969, 434 pp.) makes the following illuminating comment: "According to the author's view, which I think nearly all biologists must share, the species is the only taxonomic category that has at least in more favourable examples a completely objective existence. Higher categories are all more or less a matter of opinion".

The fact that men are able to arrange plants and animals in a classification table on the basis of their morphologic features, certainly is no proof that those more closely associated in the table are more directly related by evolutionary descent. All such an arrangement proves is that man has the ability to devise methods for classifying and categorizing assemblages of data.

As a matter of fact, the classification table is a much better support for the creation model. If an evolutionary continuum existed, as the evolution model should predict, there would be no gaps, and thus it would be impossible to demark specific categories of life. Classification requires not only similarities, but differences and gaps as well, and these are much more amenable to the creation model. (Morris, 1974, pp. 71–2. The review was G. W. Richards, 'A guide to the practice of modern taxonomy', *Science* 167, 1477.)

The half-truths, or rather quarter-truths, of this whole passage are clear for all to see. Does one laugh or cry at the insinuation that Mayr thinks the divisions in the animal world are all so arbitrary that one simply cannot separate out man, horse, bat, mole, and porpoise, for it is this that one has to accept if one is to take Mayr's work as denying the evolutionary relevance and importance of the homologies between the front limbs of all of those organisms. And how, other than with contempt, can one treat the claim that evolutionism implies "there would be no gaps between living species"? Certainly one can have an evolutionary theory with such an implication. Indeed, we ourselves shall encounter one shortly. But the synthetic theory, based as it is on a branching picture of evolution, presupposes gaps in the most fundamental way possible, and as we know, expends much effort on explaining the process of speciation, the key cause of such gaps.

I will take no more time with the scientific pretensions of Special Creationism. I find them self-servingly naive to the point of dishonesty. I do not think that principles of free-speech demand that such ideas be taught in the classroom, and most certainly not in the time allotted to science. They seem to me to be on a par with claims that it is permissible to persecute Jews because they crucified Our Saviour. Furthermore, let me state that if God does exist and did create the world in the way the Special Creationists claim, then He seems to have been peculiarly inept in using the muddled and often contradictory books of the Old Testament as His medium of communication. Indeed, He seems to me to have been somewhat sly (or a bungler) in leaving around such phenomena as Darwin's finches on the Galapagos, which speak so strongly of an evolutionary origin. Why did He give us our reason if it was going to mislead us so? I want no part in such a God as this, and, if refusal to take Him seriously implies that I will go to hell, then, to quote a far greater philosopher than I, "to hell I will go" (Mill, 1865).

2.4. RIVALS: LAMARCKISM

I move on now to evolutionary alternatives to the synthetic theory. The first alternative is the most famous of them all: *Lamarckism* (Lamarck, 1809). There is, in fact, some ambiguity about the exact nature of this position, so-named after the French evolutionary thinker Jean Baptiste de Lamarck (Burkhardt, 1977; Russell, 1916; Mayr, 1973; Hodge, 1971). Today, by 'Lamarckism' we refer to the inheritance of acquired characteristics and any related causal mechanisms — phenomena like the blacksmith's arms getting developed through striking at the anvil all day, which developed arms are then passed directly to his children, who are born so endowed (Cannon, 1958). Popularly cited examples of Lamarckism are the thick soles on the feet of humans, the callouses on the rear end of ostriches, and, of course, the neck of the giraffe (Figure 2.5). However, for Lamarck himself the inheritance of acquired characteristics was but a small part of his theory. Indeed, it was a side mechanism and it was certainly not original with him. Paradoxically, it played at least as important a part in a work by a later evolutionist, *On the Origin of Species* by one Charles Darwin (Ruse, 1979a). Lamarck's main theory had, at its centre, something very much like an escalator. He believed that small organisms like worms are constantly being formed by electrical action out of mud and dirt, and then through a kind of internal teleological force brought about by environment needs, they and their successors start a slow progressive climb up the 'chain of being' — fish, reptiles, amphibians, mammals, humans (Lamarck, 1809; plants have a separate chain). The key assumption is that new organisms are constantly getting on the ladder or escalator of life at the bottom and then, with a few side variations, Lamarck believed that there is an inevitable succession up through a pre-determined route. Thus, if all elephants were destroyed tomorrow, they would re-evolve.

One obvious consequence of 'true' Lamarckism is that the organisms we see around us today do not have common ancestors (Figure 2.6). This means that we should not expect to find evidence of such ancestry in the fossil record! Interestingly, although perhaps not so surprisingly, Lamarck himself was basically indifferent towards the fossil record — he certainly did not use it as support for evolutionism. Of course, gaps between types or organisms (in the fossil record) were a problem for Lamarck: temporarily speaking he

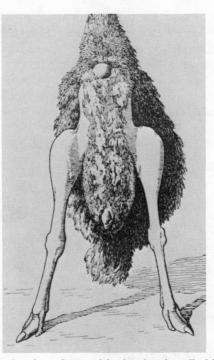

Fig. 2.5. View of ventral surface of an ostrich, showing the callosities (from Waddington, 1957, p. 161).

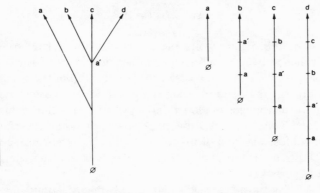

Fig. 2.6. Difference between a theory like Lamarck's and one of common descent. Life is supposed to begin at ∅, and *a, b, c, d*, are kinds of organisms extant today (from Ruse, 1979a).

thought there were no gaps. But more importantly, given constant creation of new life ('spontaneous generation'), Lamarck thought there ought to be no gaps between organisms as we see them today. Fossil gaps he tended to ignore or deny, and he took much the same approach to gaps which we find between today's organisms, say between horses and cows. Any such gaps Lamarck explained as a result of our not yet having discovered the missing transitional organisms, or possibly as having perished due to human intervention and destruction. The story of the dodo, the flightless South Sea island bird that humans drove to extinction, was well known (Figure 2.7).

Fig. 2.7. The Dodo.

On top of this organic-world picture, particularly in later versions of his theory, Lamarck laid the inheritance of acquired characteristics. This latter mechanism, he thought, would lead to deviations from the straight upward progress through the chain of being and could, in fact, go so far as to result in a kind of branching evolutionary effect (Figure 2.8). But it must be noted that Lamarck's was still not a theory of common descent. One should, therefore, not confuse the kind of branching picture that Darwinian evolutionists give to illustrate their position (Figure 2.9) with the superficially similar picture of Lamarck.

No one today takes Lamarck's original main theory seriously, nor do living biologists accept the inheritance of acquired characteristics. This is not because of dismissal, 'possibly not on very adequate evidence', but because absolutely massive evidence shows these twin hypotheses to be quite false. The fossil

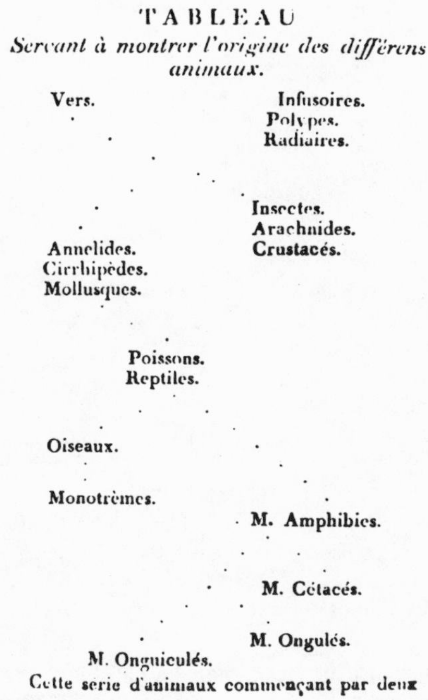

TABLEAU

Servant à montrer l'origine des différens animaux.

Vers.   Infusoires.
         Polypes.
         Radiaires.

        Insectes.
        Arachnides.
Annelides.   Crustacés.
Cirrhipèdes.
Mollusques.

    Poissons.
    Reptiles.

Oiseaux.

Monotrèmes.

      M. Amphibies.

      M. Cétacés.

      M. Ongulés.
M. Onguiculés.
Cette serie d'animaux commençant par deux

Fig. 2.8. Lamarck's picture of evolution, from his *Philosophie Zoologique.*

record shows no trace whatsoever of the kind of escalator evolution proposed by Lamarck. To the contrary, if an organism lived and flourished at one period, and then dies out, it never reappears at another period. In addition, all our knowledge of genetics suggests that new characteristics do not appear to order as needed — most mutations are deleterious (Dobzhansky, 1951). There is certainly no evidence that two separate populations of the same species will evolve down the same path, as Lamarck supposes they must. Darwin's finches and like phenomena suggest the opposite. And no longer can we explain away gaps between species as simple failure to find the linking members, or even as victim's of human violence. The gaps are real. (I am referring here to gaps between living organisms, like that between pigs and sheep. Lamarck denies the reality of these except inasmuch as one might get the odd case caused by his secondary mechanism.)

What about the concept of spontaneous generation, so critical to Lamarck's theory? In fact, it has a rather interesting history, although I fear not one

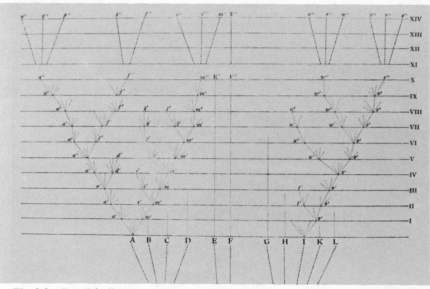

Fig. 2.9. Darwin's diagram of descent in the *Origin of Species*. Times moves upwards.

which brings much comfort to a neo-Lamarckian (Farley, 1977). By the mid-nineteenth century it was apparent to all but the most credulous, that spontaneous generation as popularly conceived, specifically as conceived by Lamarck, was simply false. All the putative cases of spontaneous generation proved to have other, more-likely causes — for example, one celebrated instance of the supposed electrical production of mites turned out to be due to the dirty fingers of the experimentors! (Chambers, 1844; Sedgwick, 1845). Although the work of Pasteur put the final nail in that coffin, even before this, in 1859, in the *Origin*, we find that street-wise scientist Darwin very careful not to saddle his speculations with spontaneous generation. There are a few vague references to an initial one or 'a few forms' (Darwin, 1859, p. 490), and that is it.

However, in recent years, thanks to some rather exciting experiments showing that essential building blocks for living beings can be created from mixtures of inorganic materials by electrical shocks, the natural creation of the organic from the inorganic has made somewhat of a come-back. At least — a sure sign — text-books on evolutionary theory now have discussions of the matter! (See, for instance, Dobzhansky *et al*. (1977) and Figure 2.10). But there is no question of Lamarck being re-established. All the recent work,

PREBIOTIC ORGANIC COMPOUNDS

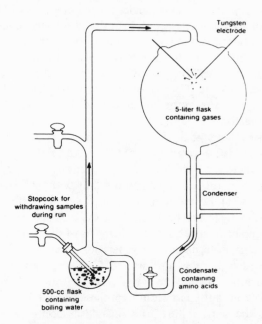

Fig. 2.10. Diagram of the apparatus used in experiments attempting to simulate the origin of life, or at least of life-supporting macromolecules. A mixture of methane, hydrogen, and ammonia, together with water-vapour, was circulated and subjected to electric shocks. It was found that from these substances, all of which presumably existed on earth before life ('prebiotic'), certain macromolecules called 'amino acids' were formed. As we shall see in Essay 5, these amino acids are a vital part of living things (Diagram from Dobzhansky *et al.*, 1977, p. 355).

speculative at best, assumes that once life got under way, no new forms would be created. If nothing else, life itself would alter and destroy the required conditions. Parenthetically, I might note an understandable reluctance by modern evolutionists to the use of the term 'spontaneous generation' to characterize today's hypothesized life-creating processes. But, although I am sure that the chief motivation behind this linguistic purity is the desire not to be tainted by what is seen as an unfortunate episode in biology's history, there is good scientific reason not to resurrect the old term. For Lamarck, spontaneous generation involved the one-step creation of fully-formed organisms like worms from inorganic matter. Today, at best, the natural creation of life is thought to be a very gradual process, involving many steps.

Lamarck's subsidiary hypothesis, the inheritance of acquired characteristics, still finds supporters in some circles today (Koestler, 1971). These circles lie outside biology because the cytological/physiological counter-evidence is decisive. For 'Lamarckism' to work, influences on the physical body must somehow get distilled into the sex-cells — else how can the effects of such influences be transmitted? But it is now known that there is simply no way in which the sex-cells can be affected by changes in the physical body (Churchill, 1968; Darlington, 1966). In the human female, for instance, the precursors of the sex cells are formed before birth. Nor is there any evidence that parts from the physical body go down (or up) to the sex cells, as the inheritance of acquired characteristics requires. That avid Lamarckian, Charles Darwin, supposed that there are little gemmules given off by the body, which then go to the sex cells; but this is a supposition totally without foundation (Darwin, 1868; Provine, 1971). Furthermore, contrary to Lamarckism, if every member (or every male or female member) of a species is subject to some environmental stress, it make no difference to future members of the species. Jews have been circumcising their male babies for generations, yet every little Jewish boy is still born with a foreskin (Mivart, 1870). Sometimes it is objected that Lamarckism is intended only to cover responses to needs and that this excludes circumcision, but apart from hypothesized benefits from circumcision (cleanliness?), the same failure of Lamarckism occurs in things involving needs. The children of blacksmiths are not born with stronger arms than the children of others, except insofar as there has been a selective process for strong men to become blacksmiths in the first place. Remember: the children of Englishmen do not innately speak English, the children of Frenchmen do not innately speak French, and the children of priests do not innately speak Latin.

Every now and then it is reported, particularly in the popular press, that Lamarckism has been 'proved'. However, closer examination shows invariably that, at best, the reference is to a phenomenon mimicking Lamarckism. One of the most fascinating of these quasi-Lamarckian effects was achieved by the geneticist C. H. Waddington (1957). He found that certain fruitflies, if subjected to heat-treatment during development, develop distinctive wing-deformities. By breeding from these deformed flies, Waddington was eventually able to produce flies with distinctive wing-deformities, even though they had had no heat-treatment! Lamarckism proved? I am afraid not, for Waddington was able to supply a perfectly orthodox Mendelian explanation. It is known that some phenotypic effects are controlled by more than one gene (i.e., genes at more than one locus). When one has but one or a few such genes, the phenotypic effects does not occur unless one has a special environment input. But if one has several of these 'polygenes' the effect occurs, whether the environmental input is there or not. Perhaps being overweight is one such effect controlled by polygenes. Some people require a special input to go overweight, namely too much food. Others, however, go overweight on even moderate amounts of food.

Waddington hypothesized that wing-deformity might be controlled in a similar way and that, in fact, what he was doing in his experiment was selecting organisms which his heat-treatment revealed as having one or a few of the pertinent polygenes: that is, the selected organisms had enough polygenes to cause wing-deformity with input, but not enough polygenes for wing deformity without input. Eventually in his experiment he would have organisms carrying polygenes in a sufficient number that no input would be necessary. If this hypothesis was true, Waddington predicted that one should occasionally find that naturally (i.e., without selection), by chance a fly will collect enough pertinent polygenes to cause wing-deformity without special environment input, and indeed this prediction proved true. In short, no special Lamarckian explanation was required. One might add that there was an epilogue to this story, namely that Waddington did think that pseudo-Lamarckian effects of this kind, which he labelled 'genetic assimilation', are important in the evolutionary process. But evolutionists have not agreed with him: critics have pointed out that, oceanic islands notwithstanding, wing-deformities are not the sort of things which usually help in the struggle to survive and reproduce (Williams, 1966). Hence, genetic assimilation is but a footnote in the evolutionary story.[2]

2.5. RIVALS: SALTATIONISM

The second evolutionary alternative to orthodox neo-Darwinism that must be considered is some variant or other of the saltationist position. This approach, one which found favour after the *Origin*, both amongst Darwin's friends and amongst his foes and which is still well-regarded today in some circles, accepts that natural selection can bring about some change, but argues additionally that major evolutionary changes require special kinds of instantaneous causes: in today's terminology, major changes require macromutations (Hull, 1973a; Ruse, 1979a). As one of Darwin's critics put matters, although he could accept selection of small variations for small changes, "where a generic limit has to be passed, bearing in mind how *persistent* generic differences are, I think we require a *saltus* (it may be a small one) or a real break in the chain, namely, a sudden divarication" (unpublished letter from W. H. Harvey to C. Darwin, 24 August, 1860). He drew an analogy with the effects of a kaleidoscope: as one turns it one gets small changes and then, in a sudden, the whole picture changes.

Somewhat paradoxically, the nineteenth-century saltationists felt they could and, indeed must, propose macro-variations for one of two diammetric-ally opposed reasons. Some, like Darwin's bulldog T. H. Huxley, were rela-tively unimpressed by the adaptive, integrated nature of animals and plants (Huxley, 1854—8). For them, things like the survival/reproductive value of the camouflage patterns of butterflies' wings were not really important facets of the organic world. Hence, they felt no pressing need for evolutionary mechanisms which would preserve or create such adaptations (i.e., mechanisms like natural selection). Therefore, when they sensed that evolution needed a bit of a push to speed things up (especially in the face of criticisms from the physical sciences that the total earth-span was far too short for a leisurely process like selection), they felt no compunction about calling on large changes — changes taking organisms from one species to another at the very least, if not greater changes (Burchfield, 1975). On the other hand, there were those thinkers, generally of a religious frame of mind, who were so imbued with the adapted, functional nature of the organic world — usually as a result of years of contemplating the argument from design — that they simply could not believe that any 'blind' lawbound mechanism like selection could suffice to produce organisms, animal and vegetable. The hand and the eye had to have something more, and this 'something more' was large variations which would create new kinds of organisms, complete with adaptations, in one generation (Mivart, 1870).

Saltationism in one form or another lasted on and off right into this century, with scientific supporters to be found in the 1950s at least (Gold-schmidt, 1940, 1952; Schindewolf, 1950). There was a tendency for the early Mendelians to be saltationists: somewhat naturally, in the early stages of the development of their theory they concentrated on large, easily identifiable mutations, and this led readily to the belief that it is these large mutations which are the stuff of evolution (Provine, 1971). It was only as work pro-gressed that there was a swing to the realization that it is small variations that really count, as they are preserved and fashioned by selection. Interestingly, among the die-hard defenders of saltationism, we find the same split as in the nineteenth century over variation and organization. The paleontologist, O. Schindewolf, impressed by the gaps in the fossil record and the sudden appearance of new forms, argued for non-adaptive macromutations. He did not see the basic groundplan of different kinds of organisms (e.g., the six-leggedness of insects) as having adaptive significance and, thus, he felt able to argue for non-directed large variations (Schindewolf, 1950; Grene, 1958). On the other hand, the geneticist R. B. Goldschmidt was so struck by the nature and widespread extent of organic functionality that he could not see how selection and small variation could possibly be the chief causal source and he, therefore, felt compelled to plump for large directed (or as-if-directed) changes (Goldschmidt, 1952).

There are at least three reasons why today's evolutionists are not salt-ationists. First, there are no known genetic or cytological mechanisms for saltations, and all the evidence we have in this direction is against them. (I shall qualify this stark claim shortly). If major mutations do indeed occur – mutations which presumably are capable of taking organisms at a minimum across species, if not a great deal further – then one would like some information on their nature and existence. All the evidence that we have so far is that the greater the mutation (or rather, the greater the effect of the mutation), the more disruptive it is on the functioning of the organism (Dobzhansky, 1951). Of course, failure to find something does not mean absolutely that the things does not exist, but there surely comes a time when continued belief is not that reasonable. No one as yet has seen a unicorn, but is it reasonable to believe in their existence?

The second reason follows on the first. If, indeed, major evolutionary changes follow on macromutations, then in sexual organisms, one is going to need at least two and perhaps a number of such mutations at virtually the same time. The whole point of these mutations is that supposedly they carry one across a species barrier, if not a great deal more. But if this is so, then a macromutated new organism cannot breed with an organism from the parent population. Hence, it must breed with a like-macromutated organism. This means that one must suppose not just one, unseen, unknown large variation, but a number of such variations. If one is improbable, how much less probable are two or three? (Dobzhansky, 1951).

The third reason against such saltations is that, if they occur and are important, then presumably one ought not see species in the gradual making. A saltation takes organisms from one form immediately right across to another. As one of Darwin's contemporaries, the astronomer/philosopher John F. W. Herschel put matters, saltationism incorporates the "idea of Jumps ... as if for instance a wolf should at some epoch of lapine history take to occasionally littering a dog or a fox among her cubs" (unpublished letter from Herschel to Charles Lyell, 14 April, 1863). But, in fact, in nature we

find just about every grade of species-in-the-making that one could imagine.
Thus, take, for instance, certain S. American species of fruitflies. We have six
closely related species of *Drosophila*, as shown in Figure 2.11. (The species

Fig. 2.11. Geographical distribution of six closely related species of *Drosophila*, with
D. willistoni and *D. equinoxialis* divided into two subspecies, which are already on the
way to separate species (from Ayala and Valentine 1979, p. 208).

are so-called 'sibling species', meaning that the members all look more or less
alike.) The sub-species of *Drosophila willistoni* are in the first stages of speciation because in the laboratory we get the following matings:

> ♀ *D.w. willistoni* X ♂ *D.w. quechua*: fertile female and male prog-
> eny.
> ♀ *D.w. quechua* X ♂ *D.w. willistoni*: fertile female but sterile males.

Already certain barriers are going up, and they are even greater between
Drosophila equinoxialis sub-species, because all crosses of *D.e. equinoxialis*
and *D.e. caribbensis* produce sterile males (and fertile females). Then, as we
move on to look at the semi-species of *Drosophila paulistorum*, we find
the speciation process yet further down the road (Figure 2.12). Where

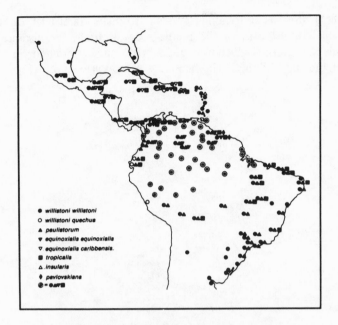

Fig. 2.12. Geographic distribution of the six semispecies of *Drosophila paulistorum* (from Ayala and Valentine, 1979, p. 211).

semi-species are geographically separate, laboratory tests show partial sterility (males not females). However, where the semi-species occur together, the members of the two groups refuse to cross-breed. There is, in fact, ecological isolation, showing that in these cases speciation has *de facto* occurred. Finally, going back to the *D. willistoni* group (*D. willistoni, D. equinoxialis,* and *D. tropicalis*), we find no natural interbreeding between members of the group and rare laboratory hybrids are totally sterile. Speciation is complete, and it seems improbable in the extreme that macromutations were involved because the *D. willistoni* are simply the end-point of a series of groups with gradual, ever-increasing, sterility barriers (Ayala and Valentine, 1979, pp. 207–12).

Perhaps the saltationist will argue that jumps occur only at higher levels, for instance, in the evolutionary transition that all of today's evidence, like embryology and comparative anatomy, leads one to think must have occurred taking reptiles to birds. This could mean that speciation normally involves only small changes. However, whilst it is indeed true that the fossil record does often show abrupt changes from one form to another (more on this in a moment), there is still enough evidence to make traditional saltationism improbable. In the case of the reptile-bird transition, for instance, there are fossils of organisms which were undoubtedly somewhere between full reptilehood and full birdhood. The famous *Archaeopteryx* is one (Figure 2.13).

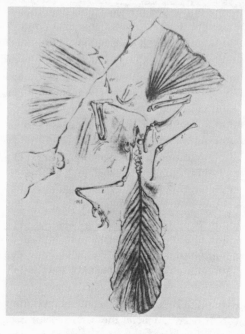

Fig. 2.13. These are the remains of the first-discovered reptile-like bird, Archaeopteryx, unearthed in Bavaria in 1861.

Moreover, there is increasing evidence of a continuous path over what is surely one of the greatest changes of them all, apes to humans (Washburn, 1978). And apart from all of this, surely one must agree that the saltationist position is starting to look increasingly *ad hoc*. If known processes are apparently adequate for small and moderate changes, why invoke unknown processes for large changes, when all the evidence is that the larger the mutation, the more deleterious is the effect?

Yet, in concluding this discussion of saltationism, two points must be allowed — not so much by way of concession, but more by way of qualification. No scientist believes any longer in the old-fashioned macromutations. Nevertheless, there are perhaps two kinds of phenomena which are saltationary

in a sense. First, it is well-known and documented that, in certain circumstances, it is possible for distinctive changes to occur at the level of the
chromosomes — sometimes instead of single haploid sets being passed on by
each parent, one gets more than one such set transmitted (Stebbins, 1950;
Dobzhansky, 1970). Offspring, therefore, get three or four haploid sets
(rather than the usual two), sometimes from just a single parent species and
sometimes from different species. What is worthy of note about this process,
'polyploidy', is that the newly-formed offspring often cannot breed with
their parent's(s') population(s) and, thus, they form new species. Polyploidy,
something which incidentally can be brought about at will in the laboratory,
is rare in the animal world: apparently it causes an imbalance between the
sex chromosomes and other chromosomes. However, there is good evidence
that polyploidy has been a major cause of evolution in the plants — 47%
of all flowering plants (angiosperms) are polyploids. All of this is certainly
not saltationism in the sense understood by the early or later supporters of
saltationism — it is unknown among the higher animals which were the major
concern of the saltationists — but it is, nevertheless, an important part of the
evolutionary picture.

Second, to complete the story it should be recorded that today a growing
number of paleontologists are arguing that the history of life, as revealed
through the fossil record, is not really one of slow uniformly gradual change
and speciation, from one or a few forms: the sort of organic 'tree of life'
that one sees illustrated in many works on evolution (Figure 2.14). Rather,
as acknowledged above, they argue that new forms frequently do appear
abruptly in the fossil record, which forms often then persist in a relatively
unchanged state through many ages. Nor are these paleontologists prepared
always to explain away the sudden coming of new forms as illusions caused
by the imperfections of the fossil record: "The paleontologist's gut-reaction
is to view almost any anomaly as an artifact imposed by our institutional
millstone — an imperfect fossil record . . . [But] we suspect that this record is
much better (or at least much richer in optimal cases) than tradition dictates"
(Eldredge and Gould, 1972, p. 97). In other words, it is claimed — on the
basis of the fossil record — that much important evolutionary change is
'saltationary'.

Nevertheless, before the traditional saltationist triumphantly claims
victory, it must be pointed out at once that these neo-saltationists do not
reject the synthetic theory of evolution. To the contrary: they state explicitly
that their position is one dictated by the theory, and is given causal under-
pinning by modern population genetics! It is argued that the hypothesis of

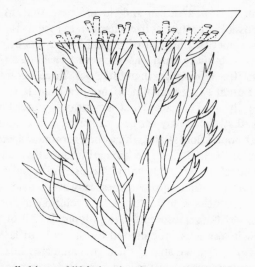

Fig. 2.14. The so-called 'tree of life', showing the supposedly gradual nature of evolution (from Eldredge and Gould, 1972, p. 109, reproducing the figure 637 given in J. M. Weller, *The Course of Evolution*, McGraw-Hill, New York (1969)).

'punctuated equlibria' is just what one would expect if the modern beliefs about speciation hold, namely that new species are frequently started by a few founders, separated geographically from their parents, and initially undergoing fairly drastic genetic change as the limited range of alleles sort themselves out into a functional, integrated gene-pool (i.e., corporate genotype). In other words, what one gets is rapid evolution as organisms go across the species barrier (and possibly across even greater barriers), followed in successful cases by periods of relative stability. And this, it is argued, is reflected in the abrupt nature of the fossil record: this picture of a 'saltationary' life history "merely represents the application to the fossil record of the dominant theory of speciation in modern evolutionary thought" (Eldredge and Gould, 1972, p. 108). Far from breaking with neo-Darwinism, paleontologists who argue this way, feel that for the first time they are truly bringing the theory to bear on their subject-matter.

2.6. RIVALS: ORTHOGENESIS

Our third and final evolutionary alternative to the synthetic theory is so-called *orthogenesis*. It was argued, particularly by a number of paleontologists towards the end of the last century, that the fossil record shows certain trends, for instance, from smaller to larger body-size in many groups of animals, and that these trends cannot be explained adequately by any theory based on natural selection (Simpson, 1953). Apart from anything else, argued the orthogenetic critics, towards the ends of the evolutionary trends, features evolve to such a state that they are positively non-adaptive. The most popular example of such a trend which had, as it were, gone over the edge, was that which culminated in the truly colossal antlers of the so-called 'Irish Elk'. Here, it was argued, was a phenomenon which could not possibly have been caused by natural selection. (The 'Elk' is really a kind of cervine deer.)

We can deal quickly with this rival to the synthetic theory. Today its status is somewhere on a par with hypotheses about a flat earth. Apart from the total absence of any mechanism to cause such orthogenetic trends, the position flounders on two facts. First, such trends are frequently illusory. As has been pointed out just above, the fossil record is often one of rapid or sudden appearance of new forms, followed by ages of comparative stability. About the most famous of all supposed trends, reduction in the side-toes of horses, G. G. Simpson writes as follows: "There was no such trend in any line of Equidae. Instead there was a sequence of rather rapid transitions from one adaptive type of foot mechanism to another. Once established, each type fluctuated in the various lines or showed certain changes of proportion related to the sizes of the animals, but had no defined trend" (Simpson, 1953, p. 263).

Second, such trends as there are and such odd features as appear at the end of trends simply are not such as to worry the synthetic theorist. The most common trends, for instance, those taking organisms from small body size to large body size, have good adaptive backing: "Larger body size has many advantages: strength in capturing prey, fleetness in evading predators, and success in mating competition are examples. So long as increasing size does not entail offsetting disadvantages, such as an inordinate food requirement, it might be favoured by selection" (Dobzhansky et al., 1977, p. 245).

As far as the antlers of the Irish Elk are concerned, there is reason to believe that, in cervine deer, antler size is a (logarithmic) function of body size, and that consequently as body size increases so also would antler size, either as a direct effect of overall body size increase, or quite possibly, in part, causing the increase in the rest of body size due to the great adaptive value of large antlers in such deer — cervine deer use their antlers in sexual combat and for display to promote breeding success (Gould, 1973). As can be seen from Figure 2.15, rather than being atypically non-adaptive, the Irish Elk

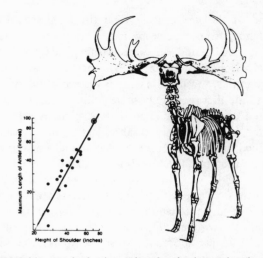

Fig. 2.15. Relation between body size and antler size in cervine deer. As can be seen, the larger the animal, the larger the antlers. The 'Elk', represented by the circled dot, has precisely the size of antler one would expect, given its body size (from Dobzhansky *et al.*, 1977, p. 244; after Gould, 1973).

is supertypical. There is no reason at all to think that the antlers were caused by forces going against natural selection. It was adaptively valuable for the deer to grow as large as possible and the antlers were part of the package deal — whether as effect, or as cause.

Orthogenetic evolution seems, therefore, to be contradicted by the fossil record — its prime evidential base in the first place — and to be unneeded in any case. If the Irish Elk is a supposedly paradigmatic instance of orthogenetic evolution, and it is, then given the way in which the animal's state and feature can be explained by the synthetic theory, orthogenetic hypotheses seem not simply false but redundant also. And with this conclusion, we can appropriately conclude our discussion of evolutionary alternatives to the synthetic theory. Lamarckism, saltationism, and orthogenesis have been tested in the fire of empirical evidence, and found wanting. The synthetic theory succeeds on its own merits, and its rivals fail on their own demerits. They had their chance and were not equal to the challenge.

2.7. EVOLUTIONARY LOGIC

In bringing this essay to a conclusion, let me make a few brief comments about the logic (or 'logic') of confirmation in evolutionary theory. I have argued that much confusion about the status of modern evolutionary thought arises from misconceptions about the synthetic theory — what it does and does not do, what it ought and need not do. Whilst it is indeed true that no one can predict the future evolutionary fate of the giraffe's neck or the elephant's nose, it is not true that the fate of the synthetic theory depends on the ability to make such predictions. The theory is concerned with processes and mechanisms, rather than simply tracing evolutionary paths. Hence, one can legitimately (and indeed must) consider processes as they are going on here today, rather than simply speculating over long-time actual routes (phylogenies). Admittedly, the synthetic theory applies (or wants to apply) to organisms like giraffes and elephants, not to mention long-gone organisms like dinosaurs. But, given the practical difficulties and impossibilities of full study of these organisms, it is quite fair that evolutionists should study manageable, fast-breeding organisms like fruitflies. For all the differences, there are significant biological similarities between fruitflies and giraffes and elephants. Therefore, having studied fruitflies, evolutionists can justifiably argue analogically to slower-breeding, less manageable organisms — or even to organisms long gone. Despite what critics sometime argue, analogy is a legitimate form of scientific inference (Salmon, 1973). And it is used by physicists. Remember that Kepler first worked out the elliptical orbit of Mars, feeling free then to generalize to other planets (Hanson, 1958). Was it wrong of him to do so?

But what about Himmelfarb's charge that evolutionists follow Darwin in wrongly taking possibilities to add up to probabilities? (Himmelfarb, 1962). Is she right in suggesting that the logic of evolutionary confirmation is wrong — at worst evolutionists are dishonest, at best they are self-servingly naive? One defence is certainly not open to the evolutionist. I have been at pains to emphasize that the synthetic theory is a little bit like a flan, with a core (population genetics) throwing light on subsidiary disciplines (like biogeography), where the various parts come to make a whole, and where indeed the whole is considered greater than the sum of the parts. (Refer back to Figure 1.3.) For instance, the paleontologists' recent speculations about 'saltation' do not stand alone, justified only by the fossil record. They are made plausible by what we know of population genetics, and also indirectly by the fact that they mesh with the work of those evolutionists working in other areas. The Galapagos finches, in a sense, help to bolster the paleontologists; and conversely the paleontologists support the biogeographers. Himmelfarb is right, therefore, in thinking that we do have possibilities adding to probabilities, and more. The evolutionist cannot escape her charge by denying that, in evolutionary theory, the whole is certainly intended to be more than the sum of the parts.

Where Himmelfarb is wrong is in thinking that this logic is the logic of chance, as in the chance of throwing two sixes given that there is a one in six chance of throwing one six. In evolutionary studies we are not dealing with logically distinct items, where the truth of one item has no effect on the truth of any other. The appropriate model for evolutionary thought is rather the logic of the courtroom, where in the absence of direct evidence, one builds a case circumstantially. Consider the sad story of Lord Rake, found dead in his study with a knife through his heart. No one saw Lord Rake stabbed, but we pin the guilt on the butler because of a number of indirect facts. The stabbing was done by a left-handed man and the butler is left-handed. The butler's alibi was false. The butler had motive; Lord Rake seduced his daughter. The butler's shirt is blood-stained. And so forth. Individually none of these facts may be overwhelming. Together they make a case strong enough for conviction. Similarly for the synthetic theory of evolution. The parts come together to make a whole. Although much work yet remains to be done, today we can confidently say that we know in outline why and how organic evolution occurs. We have hanged men with less evidence.

NOTES

[1] I can speak to this myself. Recently I published a book on the Darwinian Revolution which, although sympathetic to Darwin and his followers, tried also to see some merits in the opposing forces (Ruse, 1979a). One reviewer stated flatly that he found my approach 'offensive' (Ghiselin, 1980).

[2] No sooner had I penned these words than my claim that reports of Lamarckism's vindication keep reappearing was triumphantly (depressingly?) confirmed. Two Canadian researchers believe that they can produce effects in parents which can then be transferred directly to offspring. However, even if these claims stand the test of time (and a great many such claims have not), what I have written in my main text will need little or no modification. The supposed effects are rather special, involving immunological responses, and certainly do not support traditional Lamarckism. If there are still any blacksmiths, they will, I am afraid, have to do their own muscle building – they cannot rely on their fathers. For details see E. J. Steele, *Somatic Selection and Adaptive Evolution* (Toronto: Williams and Wallace, 1979). But see also, Anon, 'Too soon for the rehabilitation of Lamarck', *Nature* 289 (1981), 631–2.

KARL POPPER AND EVOLUTIONARY BIOLOGY

Scientists love to measure and order things, including their fellow scientists (Cole and Cole, 1973). They have elaborate and full systems of honours to acknowledge rank and importance, from Nobel Prizes down through membership in prestigious bodies (like the Royal Society) to medals awarded by small specialist organizations (the Douglas Medal awarded by the American Society for Testing and Materials). Informally also, when two or three scientists get together one often gets lists of the 'Top Ten' in the field. My own favourite is one complied in 1851 by the young T. H. Huxley, nearly ten years before the *Origin* and, indeed, before he had become an intimate of Darwin (Huxley, 1900, 1, p. 94). Heading the list was, of all people, the man with whom Huxley was to clash in such dramatic fashion in 1860 at the British Association's debate on human origins, Richard Owen! Darwin, I am afraid, came very much down the list – although to do a justice to Huxley, which he would probably not appreciate, in 1851, on the basis of published work, he was quite right to rate Owen far above Darwin.

Philosophers like to pretend that they are superior to ordinary beings. They have virtually no honours or prizes. Indeed, for payment of a small fee anyone can join the American Philosophical Association, and it will be remembered that when one of their number, Jean-Paul Sartre, was offered a Nobel Prize, he rejected it – the fact that a rejection drew far more attention than would a simple acceptance was, of course, entirely coincidental. But let me reveal a professional secret. In fact, philosophers are as given as lesser mortals like scientists to the drawing up of tables of the Great Thinkers of the past or present. I am sure few of us have gone through graduate school without wondering whether one should put Kant above Plato, and if perhaps Aristotle tops them both. My suspicion is that when philosophers of science start to play this game, there is one name which always appears on the lists, namely that of Sir Karl Popper. For some reason Popper stirs strong emotions. There are those who would unhesitatingly put Popper right at the top of their lists. Indeed, one thinker with philosophical pretensions has gone so far as to argue that Popper is the greatest philosopher of science that there has ever been, past or present (Sir Peter Medawar quoted in Magee (1974); see also Schilpp (1974)). Others view Popper less favourably and spend much time

condemning his views; but one suspects that even they would find some place
for Popper on their lists: possibly a conspicuously low one to make the point!
But do not be misled. The sign of weak philosophy is not vigorous criticism,
but indifference.

As is well known, Popper's major contribution to the philosophy of science
rests on his proposed "Criterion of Demarcation" between science and non-
science. Popper argues that the mark of true science is that it exposes itself
to the test of experience, unlike other subjects like theology and metaphysics
(Popper, 1959, 1962). Thus, for instance, a typical scientific statement — say
Kepler's claim about Mars's elliptical orbit — could be shown false. It could
be that Mars goes in squares, not ellipses. A typical statement of theology,
however — say that God loves us all — could never be shown false by experience. Even in the face of horrible diseases like leukemia, theologians find
ways to defend God's all-embracing love. And the same holds for a typical
statement of metaphysics, for example, Plato's claim that the objects of this
world of ours are imperfect copies of true reality. Nothing one can say about
our tables can prove or disprove claims about the Form of Table, Tableness-
in-itself. In short, science unlike non-science, is *falsifiable*: "I shall not require
of a scientific system that it shall be capable of being singled out, once and
for all, in a positive sense; but I shall require that its logical form shall be such
that it can be singled out, by means of empirical tests, in a negative sense: *it
must be possible for an empirical scientific system to be refuted by experience*
(Popper, 1959, pp. 40—1, his italics).

Related to this criterion of falsifiability Popper has endorsed what he
refers to as a 'Deductivist', rather than 'Inductivist', position. Often it is
thought that scientists generalize from one or a few examples, inductively.
Popper argues that this is an invalid form of inference — the only valid form
of inference when dealing with generalizations is the deductive one, whereby
a single counter-instance can show a whole general statement false. Popper,
therefore, argues that true scientific progress lies, not so much in the making
of scientific generalizations, but in the attempt to knock them down again,
to falsify them (Popper, 1962, 1972, 1975). Of course, Popper recognizes
that scientists will and must make scientific generalizations — his objection
is to the claim that the mere 'inductive' act of making general statements
guarantees these statements some sort of truth status.

Stimulating though Popper's suggestions are, his importance as a philosopher of science, and the reason why he is one of the very few of that ilk to
attract the attention and respect of real scientists, does not stem simply from
his theoretical proposals. The importance and the respect stems also from the

fact that he is able and willing to apply his ideas to actual science, showing what implications one can and should draw. Naturally, Popper has not been able to consider every area of science with equal interest and depth. It is physics which has had his most detailed treatment, followed by, I suppose, the social sciences, including those bordering on the medical—like Freudian psychoanalytic theory. Roughly speaking, I think it is true to say that Popper sees the physical sciences as doing rather well and the social sciences, particularly anything to do with Freudianism, as doing rather badly (Popper, 1962, 1974b). To date, Popper has not really had that much to say about biology, but in the past ten years he has turned his attention more and more to the life sciences. Even now he has hardly given us a fully-fledged philosophy of biology, a philosophy of evolutionary biology in particular, but it is getting to the point where we can piece together Popper's main ideas and make a preliminary evaluation (Popper, 1972, 1974a, 1975). This I shall try to do in this essay, and I think it something worth doing for three reasons.

The first reason is that, because Popper gives variants of a number of objections commonly levelled against evolutionary biology, we shall be able to complete the defence of the synthetic theory begun in the last two essays. The second reason is that already a number of biologists (good biologists) are turning to Popper's ideas to illuminate and justify methodological commitments which underly their scientific work. Before they follow physicists too far down this road, it seems worthwhile seeing what their authority figure himself has to say about their subject. (See, for instance, Ayala *et al*. (1974), or just about any article in *Systematic Zoology* in the past ten years, like Wiley (1975), Platnick and Gaffney (1977), (1978), Cracraft (1978), Nelson (1978), Patterson (1978). A recent lone scientific critic of Popper was Halstead (1980). The editor of the *New Scientist*, in which the article appeared, reported a large post-bag and, if the published replies were typical, then reaction to Halstead was almost uniformly negative. See the *New Scientist*, July 31, 1980.)

The third reason why I want to look at Popper is because recently he has been developing his overall theory of knowledge, his epistemology. He has labelled his theory of scientific theory change 'evolutionary' and he has drawn a very strong analogy between what he takes to be biological evolution and the evolution of scientific knowledge, if indeed they are not for him part and parcel of the same thing: "The theory of knowledge which I wish to propose is a largely Darwinian theory of the growth of knowledge" (Popper, 1972, p. 261). Hence, by looking at Popper on biology, tangentially it might prove possible to learn something about his more general philosophy. Given

this possibility, although my aim in this essay is basically to look at Popper
on evolutionary biology, briefly at the end I shall link up his biological views
with his more general views, evaluating the supposed connection in the light
of previously derived conclusions.

3.1. EVOLUTIONARY THEORY AS A METAPHYSICAL RESEARCH PROGRAMME

Put simply, it is Popper's claim in his writings of the past decade that, in an
important sense, neo-Darwinian evolutionary theory, the modern synthetic
theory of evolution, is not a genuine scientific theory. He argues that the
theory is not properly testable, and then, true to his most fundamental
philosophical tenets, he concludes that the theory is metaphysical. "I have
come to the conclusion that Darwinism is not a testable scientific theory but
a *metaphysical research programme* – a possible framework for testable
scientific theories" (Popper, 1974a, p. 134, his italics). One should add that
in calling the theory 'metaphysical', unlike other philosophers including
the so-called 'logical positivists', Popper is making a philosophical point
and supposedly not thereby implying condemnation. Indeed, he puts Darwinism in the same column as Deductivism (as opposed to the column with
Lamarckism and Inductivism), and there can surely be no higher Popperian
praise than that. The ostensive point therefore is not that the synthetic
theory is worthless, but rather that it does not itself tell us about how the
world is or was, and about how the world works. The theory has great value
in that it sets limits on what any satisfactory evolutionary theory will be
like, just as, for example, the laws of arithmetic set limits on what any
satisfactory theory of physics will be like. It does this much, even if it does
not do more.

Nevertheless, one might think that any evolutionary biologist would feel
his/her nose somewhat out of joint when, after years of observation in the
laboratory or the wild, he/she learns that the synthetic theory does not really
tell us anything directly about nature after all! Furthermore, one's feeling
that perhaps Popper (for all the sweet reasonableness) is not really so friendly
toward modern evolutionary thought is significantly increased when one finds
that Popper slips into referring to the synthetic theory as 'feeble', and perhaps, even more revealingly, makes a major suggestion for 'an enrichment of
Darwinism'. Could it be that Popper actually rates neo-Darwinism no more
highly than Freudianism, that is, not very highly at all?

But, however, Popper 'really' regards the synthetic theory, what truly counts, of course, are his arguments for regarding neo-Darwinism as meta-physical, or whatever. Let us, therefore, turn to these and try to assess their strength.

3.2. THE PROBLEM OF SPECIATION

Popper begins with an argument about possible life on Mars. Popper argues (correctly I think) that the Darwinian evolutionist would make at least the following three claims. First, organisms reproduce in kind fairly faithfully; second, there are small, accidental, hereditary mutations (causing change); third, there is a process of natural selection. Now, argues Popper, the evolutionist would seem to be committed to the view that if ever, on some planet, we find life satisfying the first two claims, selection will come into play and cause a wide variety or organic forms. Hence, evolutionary theory would seem to make predictions which are testable. Therefore, evolutionary theory would seem to have genuine scientific content and would seem to offer the possibility of genuine scientific explanations. However, argues Popper, Darwinian evolutionary theory does not really make such a claim about a variety of forms.

For assume that we find life on Mars consisting of exactly three species of bacteria with a genetic outfit similar to that of terrestrial species. Is Darwinism refuted? By no means. We shall say that these three species were the only forms among the many mutants which were sufficiently well adjusted to survive. And we shall say the same if there is only one species (or none). Thus Darwinism does not really *predict* the evolution of variety. It therefore cannot really *explain* it. (Popper, 1974a, p. 136, his italics.)

I make two comments about this argument. First, although evolutionists do believe that natural selection working on random variation will normally lead to variety, they do not think that this is something which must follow necessarily. As pointed out in the first essay, selection can act to keep a population or species absolutely stable by eliminating all new mutations. But this leads on to the second point. Variety will come about when and only when there is, as it were, some advantage to or cause for such variety — when different ecological niches, for example, can be used. Now this claim, it seems, does lead to predictions testable, at least in principle. Popper's Mars example is perhaps a little unfair, because we know already that Mars is not going to be very hospitable to life, so one is hardly going to expect so much organic variety as here on earth. Here on earth one gets a lot more variety in the jungle than in the desert. But let us consider for a moment some of the reasons or conditions for variety, in particular (following Popper), let us consider some of the reasons why one might get different species created.

First remembering Darwin's finches, there is the question of isolation or separation. There is, in fact, some controversy between evolutionary biologists about exactly how or why speciation occurs, and the role played by isolation — in particular, some feel that speciation (between two groups originally of the same species) always requires a period of *geographical* separation. Others, believe that, although most speciation may be of this kind, 'allopatric' speciation, it can occur between groups not geographically separated ('sympatric' separation). But even those who allow sympatric speciation often demand some kind of ecological separation — say, speciating groups being on different parts of the same host. So, separation or isolation of some sort or another seems most important for speciation. (See Mayr (1963) and Lewontin (1974) for details.)

Obviously, however, whether speciation is allopatric or sympatric, more is required. What is needed is some reason or reasons to push apart the genes of the two speciating groups. As pointed out in earlier essays, selection is clearly going to be the main thing operative here — for instance, the ecological and geographical conditions of the two groups may be very different, and these in turn might well lead to different selective pressures. And finally, let us mention or rather re-mention something which may well be of great importance in speciation, namely the so-called 'founder principle' (Mayr, 1963; Dobzhansky, 1970), based as we have seen, on the belief that founding populations of organisms will necessarily be atypical and, thus, potentially poised towards rapid initial evolution.

Now, let us start putting some of these points together in possible models. Suppose for a start, one came across a planet where the chances of allopatric speciation seemed difficult rather than otherwise — suppose, for instance, the planet were fairly small and uniformly covered with water without freakish currents, and so on. One suspects that were but a few aquatic species discovered on such a planet, no evolutionist would be desperately perturbed. On the other hand, suppose one can across a planet with conditions which seemed tailor-made for speciation. Suppose, for instance, one had an area with fairly large populations, which investigation showed to be variable genetically, which in turn manifested itself as phenotypic variation. Suppose also, on such a planet one had other isolated areas, with differing conditions — cold, warm, dry, wet, and the like. And suppose finally, there seemed possible rare (but only rare) ways in which organisms might go from the main area to the isolated areas, which isolated areas were now inhabited. Had one reason to believe that life on the planet was fairly old (e.g., through the fossil record or general complexity of structure), yet were one to find that absolutely no speciation at all had occurred, then I suggest that, *contra* Popper, modern evolutionists would be worried. Their theory, parts of it at least, would have been falsified. The claims that they make about speciation would seem not to hold.

Of course, talk of hypothetical planets, at best, makes evolutionary theory testable in principle, but this seems all that is necessary to counter Popper here, since his argument is at the hypothetical level. Nevertheless, readers of my earlier essays will know fully that there is absolutely massive empirical evidence from this world which seems both to support and test evolutionists' claims about speciation. Repeatedly through the world evolutionists have found and are finding cases where populations, isolated from the main group under the kinds of conditions described above, have evolved into new species. The classic case is that of Darwin's finches; but there are many other instances, for example, the Hawaiian Drosophila and fly catchers (Dobzhansky *et al.*, 1977). Moreover, we know that there is experimental evidence, based on populations of captive fruitflies, which supports the founder principle hypothesis in particular, apart from more general evidence supporting claims about the genetic variability always in populations, which claims are so crucial to modern thinking about speciation (Dobzhansky, 1970).

3.3. IS NATURAL SELECTION A TAUTOLOGY?

Next, in his campaign to show that evolutionary theory is metaphysical, Popper suggests that adaptation and selection are just about vacuous.

Take 'Adaptation'. At first sight natural selection appears to explain it, and in a way it does, but it is hardly a scientific way. To say that a species now living is adapted to its environment is, in fact, almost tautological. Indeed we use the terms 'adaptation' and 'selection' in such a way that we can say that, if the species were not adapted, it would have been eliminated by natural selection. Similarly, if a species has been eliminated it must have been ill adapted to the conditions. Adaptation or fitness is *defined* by modern evolutionists as survival value, and can be measured by actual success in survival: there is hardly any possibility of testing a theory as feeble as this (Popper, 1974a, p. 137, his italics).

It does not detract from Popper to note that this is the most common criticism that one hears of the synthetic theory of evolution. Repeatedly the charge is made that the synthetic theory is no genuine empirical theory because its central claim, namely that about the importance of natural selection, is simply an empirical empty tautology. Natural selection, so the critics claim, tells us no more about the real world than does a stipulative definition of 'bachelor' as 'unmarried male'. Agreement to use the word 'bachelor' in this way does not tell us whether there are any bachelors, or what they would be really like. Similarly, the term 'natural selection' tells us nothing, because it is equivalent to 'the survival and reproduction of the fittest', and since we all know that the 'fittest' are defined as 'those which survive and reproduce'', this is to say no more than that "those which survive and reproduce are those which survive and reproduce''! No doubt this conclusion is true, but it is hardly the foundation on which to build an empirical theory of evolution. Perhaps indeed therefore, characterizing the theory as 'metaphysical' is charitable. Even "God is love" seems more informative than "natural selection is at work''. (For instances, of this criticism see Manser (1965), Barker (1969), Peters (1976), although, see also Stebbins (1977), Caplan (1977).)

It is perhaps worth adding that Popper and other non-biologists are not alone in expressing these kinds of queries about natural selection. Indeed, the biological company could not be more distinguished! One of today's leading evolutionists, R. C. Lewontin, used virtually the same language as Popper at one point, even though in later writings, he backtracked somewhat (Lewontin, 1961, 1969). Surely, with all this smoke there must be some fire? There has to be something tautological or analytic surrounding selection – a definition perhaps? On the other hand, in the light of all the studies and evidence we have covered in the first two essays, it seems ludicrously implausible to suggest (at least, it does to me!) that there is nothing to evolution but a tautology. And I would argue that both of these suspicions do, in fact, have some foundation. There is a definition involved in the notion of selection as, of course, there is bound to be for any concept of science, but it is simply not the case that natural selection is a truism and that consequently the synthetic theory is empirically empty.

First, take natural selection itself. This is a systematic differential reproduction between organisms, brought about ultimately by the clash between organisms' tendency to increase in number in geometric fashion and the inevitable limitations of space and food supply. Now, in pointing to the fact that there is a differential reproduction – that not all organisms which are born survive and reproduce (offspring which are in turn viable) – we hardly have something which is tautological. The differential reproduction may be as 'obvious' as the roundness of the earth, but neither is empirically empty – certainly the differential reproduction makes evolutionary theory testable. If we all just budded off one and only one offspring asexually, evolutionary theory would be false. At least, it would be false inasmuch as one tried to apply it to this world of ours, which is precisely what evolutionists do and try to do. (In the short term, it must be allowed that one could have a differential reproduction, even though all organisms reproduce. It would just be a question of one organism or kind having more (viable) offspring than another organism or kind.)

Second, we have the point about selection which seems to cause so much trouble: natural selection means no more than the survival of those which survive. Now, it certainly seems to be the case that evolutionists do link up adaptive value and fitness in terms of survival value (or, more precisely, in terms of reproductive value), and that what we have here are definitions — analytic or tautological statements. If one says that, in places where malaria is a deadly scourge, that people with one sickling gene are 'fitter' than people without any, then we certainly do imply that the heterozygotes will have more offspring than the homozygotes. But even here we have the evolutionists doing more than just making straight analytic definitions — quite apart from having already presupposed the empirical fact that there is a differential reproduction. Evolutionists always emphasize that natural selection is *systematic* — the differential reproduction is not a random matter. Overall success is believed to be, on average, a function of organisms' peculiar characteristics, and so not only do we have the very nonanalytic matter of which characteristics aid survival and reproduction — if we were all identical there could be no selection — but also we have the claim that things of adaptive value in one situation will also be of value in similar situations. This may be difficult to test, but it is an empirical claim and could be false. Logically, it could be the case that the sickling gene protects against malaria in one case, but not another. That no such non-protective genes have been found does not prove the theory analytic — it may just be that the theory is true. Falsifiability should not be confused with being falsified. And obviously, belief in the systematic nature of selection is not simply a matter of blind faith. Evolutionists do have positive evidence, from experiments and nature, that there is a kind of uniformity about adaptive value in similar situations. For example, as we saw, winglessness seems to be of value to insects and other small animals on oceanic islands (because they stand less chance of being blown away) and, as we saw, also experiments bear this out (Dobzhansky, 1951). Similarly, there is no good reason to think that the sickling gene protects against malaria in some cases but not in others. Where malaria is active, and where the sickling gene exists, the latter protects against the former. (See Figures 3.1 and 3.2. They show dramatically the extent to which the sickle-cell

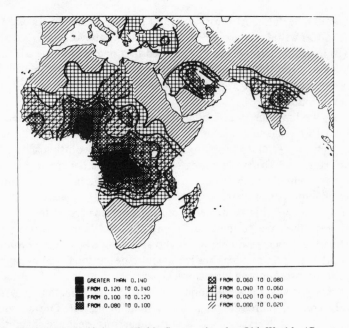

Fig. 3.1. Distribution of haemoglobin-S gene in the Old World. (Courtesy, D. E. Schreiber, IBM Research Laboratories.)

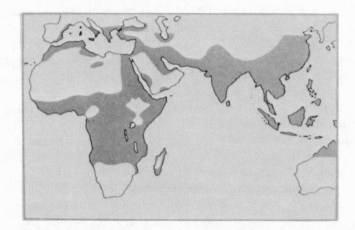

Fig. 3.2. Distribution of malaria (due to *Plasmodium falciparum*) in the Old World (from Bodmer and Cavalli-Sforza, 1976, p. 312, reproducing a map given in M. F. Boyd (ed.), *Malariology*, Saunders, Philadelphia (1949)).

gene and malaria coincide, underlining the *systematic* nature of the responsible causal process.)

Third, against Popper and fellow travellers, let us note that they ignore entirely the fact that evolutionists allow that it can be the less well-adapted which can survive and the more well-adapted which fails to survive. For a start, it is percentages which count, not individuals — does one group on average have a better record than another, not does one individual survive rather than another. Then, there is the hypothesis of so-called 'genetic drift'. It is argued by many modern evolutionists that, in certain special situations, fortuitously the less well-adapted (or neutral) can succeed where the more well-adapted gets eliminated. This effect supposedly comes about because of errors of sampling. If one has a very small population, then Hardy—Weinberg equilibrium may not hold. Suppose one had only two individuals, both A_1A_2 heterozygotes. In a large population both A_1 and A_2 would persist but, by chance, in so small a group in the next generation it might be that only A_1A_1 individuals are born. And, in a somewhat larger population, the same effect could come about in a generation or two, with one allele or another 'drifting' to total superiority or being eliminated entirely (Wright, 1931; Dobzhansky, 1951). It must be admitted that genetic drift is still a highly controversial issue, as we shall see in a later essay (Essay 5); but the way Popper argues, it would be ruled out as contradictory — the fittest could never be that which does not survive — whereas even its strongest critics seem to feel the need to mount empirical counterarguments. It should be added, lest it be thought that the possible existence of drift makes evolutionary theory unfalsifiable in the sense that all characteristics necessarily have an explanation — selection or drift — that no one today denies that selection fashioned major characteristics like the hand, the eye, and so on. Conversely, even were one to find that the sickling gene gave no help against malaria, it would be unconvincing to suggest that so deleterious a gene could have become established though drift. Hence, drift is not a ubiquitous escape clause against falsification (see Ruse, 1973a, p. 115).

Fourth and finally, to counter Popper's objection to natural selection, let us note that in neo-Darwinian evolutionary theory, linked with selection, although perhaps not really part of it, we have the claim that the selected characteristics will be passed on from one generation to the next. This is obviously necessary for evolution, for were there no such transmission, selection would have no effect. And the claim is clearly empirical — logically it is quite possible that the strong, sexy, or otherwise advantaged individuals always have puny, ugly, or otherwise disadvantaged offspring. Although an evolutionary theory based on selection does not necessarily have to use a theory of inheritance stemming from Mendelian or neo-Mendelian claims and findings — Darwin's did not, for example — we know that today's synthetic theory does, and inasmuch as it does it makes empirical claims and lays itself open to test and potential refutation.

All things considered, therefore, it seems ridiculous to keep claiming that evolutionary theory — either Darwin's or the modern synthetic theory, has as its heart a devastating tautology. The time has come to lay this misconception quietly to rest.

3.4. THE PROBLEM OF GRADUAL CHANGE

Finally, in considering Popper's arguments about the nature of evolutionary theory, let us look at an attack he launches based on the theory's claims about the gradual nature of change. Popper allows that the theory "certainly does *predict* that if such an evolution takes place, it will be *gradual*" (Popper, 1974a, p. 137, his italics). However, he goes on to say:

Gradualness is thus, from a logical point of view, the central prediction of the theory. (It seems to me that it is its only prediction.) Moreover, as long as changes in the genetic base of the living forms are gradual, they are — at least 'in principle' — explained by the theory; for the theory does predict the occurrence of small changes, each due to mutation. However, 'explanation in principle' is something very different from the type of explanation which we demand in physics. While we can explain a particular eclipse by predicting it, we cannot predict or explain any particular evolutionary change (except perhaps certain changes in the gene population *within* one species); all we can say is that if it is not a small change, there must have been some intermediate steps — an important suggestion for research: a research programme. (Popper, 1974a, pp. 137–8, his italics.)

I am not quite sure what to make of this argument, because it seems to me to be so unjust. Either a Darwinian evolutionary theory (past or present versions) predicts that change will be gradual, or it does not. In fact, for most cases it does, therefore we have a prediction, therefore it is testable, therefore it is not metaphysical. One may argue that the explanation is not of very much, but it is of something — although, in fact, if one looks at the matter historically, as I have already intimated, one finds that the gradualists had a terrific battle to win over the nongradualists, the saltationists (see Essay 2). Of course, the explanation is 'in principle' in the sense that, until one turns to an actual case, specific details are lacking, but this is the same as any theory until one turns to an actual case. The gradualness, at least, is no more in principle than are eclipses. One may agree with Popper that one has little more than a 'research programme', a start to explanations, not an end, but by Popper's own philosophy the programme is not metaphysical.

Incidentally, Popper seems unaware of the saltationist 'concessions' that supporters of the synthetic theory allow. For instance, he does not know that evolutionists believe that in the plant world evolution can occur nongradually in steps due to combining of complete sets of parental chromosomes in offspring (see Stebbins (1950), as well as Eldredge and Gould (1972)). Popper seems unaware also of the great amount of explanatory information there is about some actual cases of gradual evolution, filling out the details – for example, in the case of the horse (Simpson, 1951). However, let us not labour the point any further here about Popper's ignorance of modern biology.

3.5. POPPERIAN SALTATIONISM

Having now run through Popper's criticisms of Darwinian and neo-Darwinian evolutionary thought, we can conclude that they are all without foundation. Popper has not shown the synthetic theory to be metaphysical or feeble or any such thing. But let us now return the compliment to Popper and take a cool look at his own evolutionary position: specifically, at his suggestions for 'enriching' Darwinian evolutionism.

Popper distinguishes between genes controlling anatomy ('a-genes') and those controlling behaviour ('b-genes', in turn subdivided into 'p-genes', genes controlling preferences, and 's-genes', genes controlling skills), and he argues that changes in the latter sometimes prepare the way for changes in the former. He writes:

We can now say that certain environmental changes may lead to new preferences or aims (for example, because certain types of food have disappeared). The new preferences or aims may at first appear in the form of new tentative behaviour (permitted but not fixed by b-genes). In this way the animal may tentatively adjust itself to the new situation without genetic change. But this *purely behavioural* and tentative change, if successful, will amount to the adoption, or discovery, of a new ecological niche. Thus it will favour individuals whose *genetic p*-structure (that is, their instinctive preferences or "aims") more or less anticipates or fixed the new behavioural pattern of preferences. This step will prove decisive; for now those changes in the skill structure (s-structure) will be favoured which conform to the new preferences: skills for getting the preferred food, for example. I now suggest that *only after the s-structure has been changed will certain changes in the a-structure be favoured; that is, those changes in the anatomical structure which favour the new skills.* (Popper, 1974a, p. 139, his italics.)

And by way of example, having schematically put his position as $p \rightarrow s \rightarrow a$, Popper invites us to consider the case of the woodpecker and his beak: "A reasonable assumption seems to be that this specialization started with a *change in taste* (preference) for new foods which led to genetic behavioural changes, and then to new skills ... and that the anatomical changes came last" (Popper, 1974a, pp. 139–40, his italics).

I do not think it unfair to characterize Popper's position as one of 'behavioural saltationism'. He argues for major rapid changes in behaviour, which will then be followed by other effects, both at the genotypic and phenotypic levels. This, it seems to me, is getting fairly close to a saltus, and indeed I shall show shortly that probably this is the way Popper is thinking and that he would not feel too uncomfortable with my characterization of his views. But what about Popper's suggestions, whatever their characterization? An obvious first point is their very limited scope — the suggestions will apply only to half the organic world, namely animals. Plants do not show behaviour, expect in a limited and metaphorical way. Hence, either we must allow that the synthetic theory as it presently stands is sufficient for them, or we must come up with yet another mechanism: one which does apply to plants. One would think that neither of these alternatives is really that attractive for Popper. If, despite all the supposed weaknesses of things like natural selection, the synthetic theory is enough for plants, their evolution and their adaptations, why is it inadequate for animals? Alternatively, if we need another mechanism for plants, then what is it and why does it not apply to animals also? One has here a lacuna in Popper's position, and one suspects that however one fills it, some nasty questions will be raised.

Leaving plants aside and turning now to concentrate on animals, in order to get to the crux of the matter, let us grant Popper's way of speaking of genes-just-controlling-anatomy and genes-just-controlling-behaviour, even though the wide-spread existence of pleiotropy (genes with more than one function) may put such a clean categorization in jeopardy. Part of the difficulty with evaluating Popper's position is to know in what sense he is saying something original. Popper argues for behaviour preceding structure. But this is generally granted by evolutionists. Thus, for example, Mayr (whom Popper cites in support) writes as follows:

A shift into a new niche or adaptive zone requires, almost without exception, a change in behaviour ... It is very often the new habit which sets up the selection pressure that shifts the mean of the curve of structural variation. Let us assume, for instance, that a population of fish acquires the habit of eating small snails. In such a population any mutation or gene combination would be advantageous that would make the teeth stronger and flatter, facilitating the crushing of snail shells. In view of the ever present genetic variation, it is virtually a foregone conclusion that the new selection pressures (owing to the changed habit) would soon have an effect on the facilitating structure. (Mayr, 1960, p. 371; see also Mayr, 1963.)

So far, so good. But one assumes that Popper wants to say, indeed feels that he is saying, something more. The question is, precisely what? One possibility is that Popper is hypothesizing a kind of behavioural mutation pressure, which as it were, following on a non-genetically caused behaviour switch, takes organisms irreversibly from one behaviour pattern to another, with anatomy hopefully following. This certainly seems different from Mayr's position, who seems to be envisioning just a reversible switch of preference, not one dependent on or necessarily involving a whole new set of genes. Nevertheless, if this is, in fact, Popper's position, then it is hard to see precisely why Popper's hypothesis is even needed. Organisms switch behaviours and, eventually, anatomical changes follow. Why bother with sandwiching preference-and-skill behaviour-changing genes in the middle? Of course, one might perhaps get such genes, but it is difficult to see what they are doing (and nothing else is doing) that Popper feels needs doing, particularly since all they seem to be doing is cutting down an organism's options. Popper writes that (non-genetically fixed) behaviour change "will favour individuals whose *genetic p*-structure (that is, their instinctive preferences or 'aims') more or less anticipates or fixes the new behavioural pattern of preferences" (Popper, 1974a, p. 139). But it is difficult to see why this is so, unless any back sliding would be dangerous – in which case there seems just as much likelihood of a need for anatomical genetic change as behavioural genetic change.

Of course, it might just be that Popper feels that these new preference-and-skill behaviour genes will, as it were, force the organism into action – they will push the organism into new preferences and thence into new skill behaviour (or by being irreversible, force the organisms to stay with new preferences and skill behaviour), when without the genes the organisms would stay with or revert to the old preferences and skill behaviour. If this is so, and talk of 'anticipation' rather implies it, then first it is not easy to understand why (as Popper claims) natural selection would be less destructive on preference-cum-skill behavioural changes than anatomical changes. With respect to the woodpecker, Popper writes: "A bird undergoing anatomical changes in its beak and tongue without undergoing changes in its taste and skill can be expected to be eliminated quickly by natural selection, *but not the other way round*" (Popper, 1974a, p. 140, his italics). But why is this? If anything, I should have thought the opposite is the case. Certainly, a Canadian bird with a new, fixed, irreversible taste for maple syrup — a taste which it would not have or would avoid without its new genes — but with (as yet) no wood-pecker-like changes in its beak, is going to be at a distinct disadvantage com-pared to its mates, who are satisfied with more humdrum fare and who are not wasting time beating their beaks futilely against maple trees. Secondly, one would truly like some empirical evidence to support Popper's position. If it is generally true, then it seems most odd that evolutionists, either working with wild or captive populations, have found no evidence of it — and they seem not to have done so.

Of course, as before the basic problem is that Popper seems not to have the first idea about contemporary biological thought about evolutionary change, speciation, movement into new ecological niches, and so on. For example, because he ties adaptive value too tightly to survival, Popper seems not aware of the great genetic and phenotypic diversity that evolutionists see in populations, which is thought to exist for the kinds of reasons mentioned earlier (Dobzhansky *et al.*, 1977). But with this kind of diversity, a popula-tion might well carry indefinitely (say) a bird-type with beak and tongue less well-adapted than most for the usual diet. Then, when a new niche opens up, because of dispersal, a change in the ecological balance, or some such thing, the previously less well-adapted type is ready to move in and take advantage in a way barred to the bird conforming to what, hitherto, had been the best-adapted type in the group. In short, it seems that evolutionary theory has no need of Popper's suggestions.

Possibly it might be objected that I am attacking a straw Popper. I have suggested already in this section that Popper's position seems akin to a form of saltationism. Perhaps he is more of a saltationist than I have hinted, believ-ing not merely in changes brought about by normal genes (which would be subject to the kinds of queries I have raised), but also believing in changes brought about by special, major, behaviour-changing genes. There is some reason for claiming that this may be Popper's position. At one point Popper writes of a "hopeful behavioural monster" (Popper, 1972, pp. 281–4), and he portrays himself as providing a new form of the saltationary theory of R. B. Goldschmidt (1940), who as we know, supposed that important evolu-tionary changes are a function of one-step macromutations, rather than the many-step microchanges of the Darwinian selectionist. Perhaps Popper favours major behavioural changes (to be followed by structural changes), which changes come in one generation from one or a combination of gene change(s), where there is something rather distinctive about the gene change(s) involved? The trouble here, however, is that if this is indeed Popper's true position, then he counters none of the objections we have seen that evolutionists have brought against saltationary theories positing such macromutations leading to successful evolutionary 'monsters' — in particular, against Goldschmidt's theory (Dobzhansky, 1951; Ruse, 1973a). Until these objections are defused, there is no reason to treat Popperian saltationism any more kindly than other brands.

All in all therefore, Popper's 'enrichments' of evolutionary theory seem at best unneeded, and at worst wrong.

3.6. EVOLUTIONARY BIOLOGY AND EVOLUTIONARY EPISTEMOLOGY

Why? That is what we want answered. Popper is an absolutely first-class philosopher of science, who has made a real and deservedly successful effort to know and engage in actual areas of science, particularly physics. Why then should he deal in so slighting and basically unfair a way towards biology? What is it about modern evolutionary thought that leads Popper into such systematic distortions, bordering on the arrogant? Let me offer three, not necessarily mutually excluding, suggestions.

First, Popper has been working in an English or English-influenced environment for many years. Despite the fact that Darwin was an Englishman and that some of the best work in evolutionary thought since his day has been done by the English, it remains true — or at least it did until very recently — that biology is considered a very second-rate and unimportant science by the English. Anecdotally I might report that, when I was a schoolboy in the 1950s in England, although I started to specialize in science at the age of eleven, I never ever took a course in biology — it was all mathematics, physics, and chemistry. Biology was a subject for the slower students, along with geography, woodwork, and Spanish. (I am not boasting. Had I been really bright, I should have specialized in Latin and Greek.) The fact that Popper shows gross ignorance of biology and, condescendingly, feels free to move in and put it right, is a reflection of this unhappy state of affairs. One can only hope that the winds of change are blowing.

Second, my suspicion is that at least part of Popper's critical attitude towards the synthetic theory of evolution comes simply from the fact that he has not yet accepted the Darwinian Revolution. Although at one point Popper says (truly) that Darwin's great move was to show how design and purpose in the organic world can be explained in purely physical terms, almost at once Popper then goes on to say that the difficulty with Darwinism is explaining something like the eye in terms of accidental mutations, and it is clear (by his own admission) that Popper is trying to get away from the accidental in evolution, and to give some direction to change (Popper, 1972, p. 270). But this is the whole point — unless one can see that it is selection acting on 'accidents', bringing about design-like effects, one misses entirely the force of Darwinism. What Darwin tries to do above all else in the *Origin of Species* is show that 'blind' law can bring about that for which everyone else felt they had to invoke the creative interference of the Great Designer.

It is scarcely possible to avoid comparing the eye to a telescope. We know that this instrument has been perfected by the long-continued efforts of the highest human intellects; and we naturally infer that the eye has been formed by a somewhat analogous process. But may not this inference be presumptuous? Have we any right to assume that the Creator works by intellectual powers like those of man? . . . In living bodies, varia-tion will cause the slight alterations, generation will multiply them almost infinitely, and natural selection will pick out with unerring skill each improvement. Let this process go on for millions on millions of years; and during each year on millions of individuals of many kinds; and may we not believe that a living optical instrument might thus be formed as superior to one of glass, as the works of the Creator are to those of man? (Darwin, 1859, pp. 188–9.)

Somehow one feels that Popper is in a tradition which started as soon as Darwin's *Origin* appeared — a tradition which includes such men as Charles Lyell and St. George Jackson Mivart, who were evolutionists but who felt that, in order to account for the design-like effects of the organic world, one must supplement selection with other mechanisms (Ruse, 1979a). Unlike them, Popper may have no theological axe to grind; but, he seems a direct intellectual descendent. Popper's macromutations exist simply to do the work of natural selection. And this makes Popper as much a pre-Darwinian as Ptolemy was a pre-Copernican.

Third and perhaps most importantly, I believe that Popper's desire to see links between organic evolution and scientific theory growth leads him astray. This could be in at least two ways. First, consider what his theory of theory growth is (Popper, 1962, 1972, 1975). As intimated at the beginning of this essay, Popper's criterion of genuine science — that it be falsifiable — and his theory of theory growth are intimately connected. They are indeed part of the same insight. Popper sees science progressing by what he has referred to as the method of 'conjectures and refutations'. A scientist throws up a hypothesis, arrived at by some non-justifiable creative process. Then the scientist and his fellows try to knock the hypothesis down, which (logically) they can do if the hypothesis is genuinely scientific. Eventually the hypothesis is indeed falsified, and then, learning from mistakes, scientists throw up another hypothesis and we start again. Progress has been made; error has been eliminated. Copernicus may not have been right, but at least he was one step ahead of Ptolemy.

Assume that we have deliberately made it our task to live in this unknown world of ours; to adjust ourselves to it as well as we can; to take advantage of the opportunities we can find in it; and to explain it, *if* possible (we need not assume that it is), and as far as possible, with the help of laws and explanatory theories. *If we have made this our task, then there is no more rational procedure than the method of trial and error – of conjecture and refutation*: of boldly proposing theories; of trying our best to show that these are erroneous; and of accepting them tentatively if our critical efforts are unsuccessful. (Popper, 1962, p. 51, his italics.)

Now obviously this philosophical analysis – and, as before, let me emphasize that I am not being perjorative – is not a scientific theory. By Popper's own criterion of falsifiability, his method of conjectures and refutations is not open to empirical refutation. No doubt Popper does think his position descriptive of some of the best science, but essentially what Popper is offering is something which is prescriptive – it is telling us what scientists *ought* to do. The Popperian theory of scientific-theory growth is, in short, metaphysical. But this gives us an obvious clue to Popper's attitude towards Darwinian biology. If Popper can label the biological theory of evolution 'metaphysical' also, showing that it simply delimits the bounds within which biologists ought to stay, then clearly Popper has forged a strong link (if not indeed an absolutely essential link) between his own philosophy and biology. The connection between the two is not one of metaphor or illustration, but perhaps even (as he himself claims) one of identity. Hence, any reputation gained in its 100-or-so-year history by Darwinian biology (perhaps more out of England than in!) rubs off on Popper. He is just putting forward something virtually analytic anyway! How can one possibly turn him down?

Second and connected to this last point, the actual way in which Popper analyses theory growth may be influencing his approach to biology. Using 'P' for problem, 'TS' for tentative solution, 'EE' for error-elimination, Popper sees all evolutionary sequences as following this pattern:

$$P_1 \rightarrow TS \rightarrow EE \rightarrow P_2. \quad \text{(Popper, 1972, p. 243)}$$

Popper thinks that this sequence is what happened in biology. Problem: to see at all. Tentative solution: rudimentary eye. Error elimination: non-reproduction of sightless animals. New problem: to see better than fellows. Popper thinks also that this sequence is what happens in science. Problem:

to find a solution to organic origins. Tentative solution: Darwin's theory. Error elimination: fall of special creationism. New problem: solve difficulty of heredity. But the fact of the matter is that, in reality, tentative knowledge solutions are frequently fairly large (saltationary) and often designed — in the case of selection theory, Charles Darwin worked intensively for nearly two years to solve the actual problem of the mechanism of evolution and spent another twenty years putting his ideas together in a coherent theory (Ruse, 1979a). There was certainly no blind chance involved in either natural selection or the overall theory of the *Origin*. Could it be that (mistakenly) Popper is reading features of the evolution of knowledge into the evolution of organisms, and that it is for this reason that he wants to supplement biological evolutionary theory in the ways he suggests? Certainly, if Popper is confusing philosophy and biology in the way I am suggesting, then his urge for large biological mutations becomes instantly explicable.

What conclusions can we draw from all of this? Even if we suppose that my surmises are correct and my criticism of Popper's views about biological evolutionary theory are well-taken, what then does this all imply? In particular, to ask the most important question, what does it imply about Popper's philosophical theory about scientific theory growth? In one sense, obviously, not a great deal! Darwinians do not have a monopoly on the word 'evolution'; hence nothing I have argued can properly stop Popper characterizing his views as evolutionary. Nor has anything I have said proven his general philosophical theory mistaken, although this is not necessarily to say that it is true. However, I suggest my arguments do show one most important thing. No longer ought Popper to claim close ties between his philosophical evolutionary theory and biological evolutionary theory, or feel that, somehow, some of the legitimacy of the latter rubs off on the former. The relationship between the two theories is, at best, one of weak analogy. In important respects, Popperian scientific theory evolution and neo-Darwinian biological evolution are different. This means that Popper's work must stand on its own merits; to pretend otherwise would be wrong, and slightly dishonest.

THE LAST WORD ON TELEOLOGY, OR OPTIMALITY MODELS VINDICATED

In their work biologists frequently use language apparently making reference to ends or consequences or goals, as, for instance, when they say that "the *purpose* or *function* of the heart is to pump the blood", or that "the heart exists *in order to* pump the blood". This 'teleological' dimension to biology attracts a great deal of attention from philosophers, although those of us with nasty minds suspect that the essence of the attraction does not really lie in the topic itself — certainly the attraction does not lie in solving a problem in biology for its own sake. Rather, what draws philosophers towards teleology is that one has to know, or at least it is generally thought that one has to know, absolutely no biology at all! One makes a dutiful obeisance in the direction of the life sciences by noting that the heart beats in order to pump the blood and that it does not beat in order to produce heart sounds, and then one can cheerfully move the discussion into the unreal world of the imagination so beloved by philosophers. Biology forgotten, one can talk about door knobs serving as paper weights, sewing machines with buttons to be pressed when one wants them to self-destruct, multi-purposed water beds, or whatever else it is that turns one on; in the unreal world of the imagination, that is! And if someone like myself tries to bring matters back to biology, smugly pointing out that perhaps heart sounds do serve an end, namely soothing babies at the maternal breast, the objection is considered irrelevant — almost bad form in fact. Philosophers want no empirical factors deflecting them in their neo-scholastic pursuits (Figure 4.1).

This is all a bit of a pity, because I think the teleology of biology poses interesting and important philosophical questions, which are being muffled in the replies and counter-replies of many of today's writers. Somewhat presumptuously therefore, let me try to swing the discussion back to its biological roots by making my own proposals. I emphasize that what interests me is biology, particularly evolutionary biology. I do not know whether what I have to say has wider application, say in the social sciences. Perhaps it does; but I shall not risk gutting my claims of all content in order to make them applicable to everything and anything. (Having been so nasty about my fellow philosophers, let me try to redeem myself by drawing the reader's attention to two excellent, although I think wrong-headed, recent reviews of the

(No Model)

No. 556,248. SALUTING DEVICE. 2 Sheets—Sheet 1.

Patented Mar. 10, 1896.

Fig. 1.

Fig. 2.

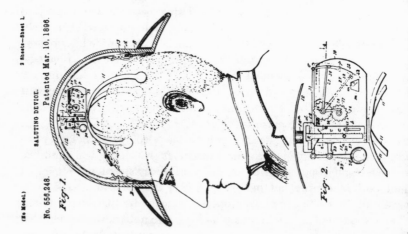

Hat-Tipping Device

UNITED STATES PATENT OFFICE

SALUTING DEVICE

Specification forming part of Letters Patent No. 556,248, dated March 10, 1896. Application filed September 18, 1895. Serial No. 562,908. (No model)

. . . This invention relates to a novel device for automatically effecting polite salutations by the elevation and rotation of the hat on the head of the saluting party when said person bows to the person or persons saluted, the actuation of the hat being produced by mechanism therein and without the use of the hands in any manner. . . .

Should the wearer of the hat having the novel mechanism within it and engaging his head, as before explained, desire to salute another party, it will only be necessary for him to bow his head to cause the weight-block 20 to swing forwardly. The swinging of the block 20, as stated, will, by the consequent vibration rearwardly of the upper end of the arm 29°, push the rod 31 backward and release the stud 34 on the rock-arm 32 from an engagement with the lifting-arm 27, so that the latter will, by stress of the spring 15, be forcibly rocked down into contact with the pin 23, as indicated by dotted lines in Fig. 2, the arm 28 having been correspondingly moved toward the lift-pin *f,* as also shown by dotted lines in the same figure. When the person making a salutation with the improvement applied to his hat resumes an erect posture after bowing, the weight 20 will swing back into a normal position, which will draw the upper end of the rock-arm 32 forwardly and more its lower end rearwardly far enough to release the arm 27 from the pin 23. The gear-wheel 23 will now be moved by the spring 15, so as to impinge the short arm 28 on the lower side of the stud *f,* which will cause the guide-plate 15 to slide upward, carrying the post 14 with it. Just before the arm 28 passes the stud *f* the detent-spring *g* will press its curved toe *g'* through the slot in the front plate of the case 10 and project said toe below the rounded lower end of the post 14. The lifting-arm 27 is now brought into contact with the pin *c,* and the pressure of the said arm on the pin *c* causes the post 14 to move upwardly in the depression *e* of the guide-plate 15 until it enters the slot *d.* The lifting-pin *c* will now be swung through the rear portion of the cross-slot *d* by the arm 27, and by the impetus given to the pin and post 14 by said arm the post, bow-piece, and hat A will receive a rotary movement sufficient to bring the pin *c* into the depression *e,* when the gravity of the parts will cause the hat to drop into its normal position on the wearer's head. . . .

83

Fig. 4.1. Functional device of a type beloved by philosophers: Saluting Device (from A. E. Brown and H. A. Jeffcott *Absolutely Mad Inventions,* New York: Dover, 1970, pp. 82–83. This project really was filed with the U.S. Patent Office.)

subject: Woodfield (1976) and Nagel (1977). A spirited, although I think even more wrong-headed, attack on the teleology problem is Wright (1976); see also Ruse (1978).)

4.1. THE TELEOLOGY OF BIOLOGY

Let us begin with an example to see the teleology of biology in action.

The stegosaurus was a large herbivorous dinosaur of the Jurassic period. What made it very distinctive was the existence of a set of bony plates running along its back (Figure 4.2). The question which paleontologists have long

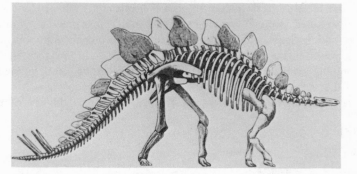

Fig. 4.2. Functional device of a type beloved by biologists: the Stegosaurus's skeleton (from Lewontin, 1978, p. 217). The original is in the American Museum of Natural History and is 18 feet long.

asked is: What 'end' or 'purpose' or 'function' did they serve? What 'problem' were they supposed to 'solve'? A number of answers were put forward. Some said that the plates existed 'in order to' make the stegosaurus seem bigger and more fearsome, thus frightening off predators. Some said that the plates existed 'in order to' facilitate courtship recognition – with plates like that down one's backside it would be pretty difficult to make a mistake and spend one's time making romantic overtures to a member of the wrong species. And some said that the plates existed 'in order to' aid heat regulation – the plates would act as sorts of cooling devices, radiating heat, thus enabling the non-sweating dinosaur to move about in the heat of the sun and get on with the business of living (Lewontin, 1978).

We see, therefore, the frankly teleological language of the biologist. And let us make no mistake about its value. By thinking in this way, biologists make great progress. Indeed, there is good reason to think that paleontologists have now solved the stegosaurus problem. The plates most probably existed for heat regulation — a conclusion based on their 'design' for such regulation:

The hypothesis that the dorsal plates of *Stegosaurus* were a heat-regulation device is based on the fact that the plates were porous and probably had a large supply of blood vessels, on their alternate placement to the left and right of the midline (suggesting cooling fins), on their large size over the most massive part of the body and on the constriction near their base, where they are closest to the heat source and would be inefficient heat radiators. (Lewontin, 1978, p. 218.)

This kind of teleological thinking is certainly not restricted to paleontology. It runs right through all kinds of evolutionary work, and indeed if anything today has received fresh impetus by the ever-increasing reliance by evolutionists on so-called 'optimality models' (Oster and Wilson, 1978). Evolutionists analyse and explain organic features in terms of problems to be solved, maximizing or optimizing certain desirable variables. Thus, for instance, in showing why praying mantises generally attack prey of a certain size, Holling (1964) argued that a prey of this size is the maximum the mantises can hold without the prey breaking loose (Figure 4.3). And the reason why the maximum-sized

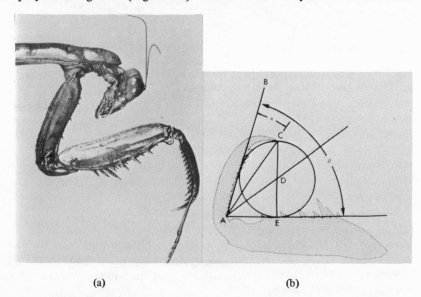

(a)   (b)

Fig. 4.3. The forelimb of the preying mantis (from Holling 1964, pp. 338 and 341). (a) is of the actual limb and (b) shows a schematic representation, demonstrating the maximum prey size that the mantis can handle, which in fact, proves to be something of diameter $T\sin(\beta-\alpha)$, where T is the length AC.

prey is to be desired is that this is most energy efficient (i.e., energy expended compared to energy consumed) from the mantis's viewpoint. Obviously this whole approach is thoroughly teleological. One is trying to understand aspects of an organism, in this case a praying mantis's forelimbs and predatory behaviour, in terms of ends, in this case optimum energy efficiency (Beatty, 1980).

What is it about biological teleology that is philosophically interesting and important? It is something which stems from comparison with physics and chemistry. The simple fact is that if one looks at the physical sciences, one just does not find the overt use of teleological language or thought or models. It is true indeed that in the last century, when Sir David Brewster — Emminent Scottish Man of Science — was asked what function the moon serves, he was happy to reply that it serves the end of lighting the way of nocturnal travellers (Brewster, 1854). But the fact that we today find such an answer amusing, points to the incongruity of using teleology in physics. The moon does not exist 'in order to' do anything. The very suggestion is not false, but absurd. It is like asking whether triangles live longer than circles. Triangles are not the sort of things that can be long or short lived, and moons are not the sorts of things that can have ends. Similarly, one can say that in chemistry things do not serve purposes. It would be quite meaningless to say that hydrogen exists in order to combine with oxygen and make water or that the function of hydrochloric acid is to combine with an alkali to make a salt.

We see therefore that, philosophically speaking, teleology (or rather, the teleological way of thinking) raises an interesting and important question. Physicists and chemists do not use it. Biologists do. What we need are answers to questions like: Why do biologists use a teleological way of thinking? How can they use it and get away with it? What do they mean when they do use it? Could biologists stop using teleological language, including optimality models, and still get as far in their biology? This last question might suggest a somewhat arrogant attitude — why on earth should biologists stop using teleological language generally and optimality models specifically? But I think it is true that biologists generally feel an urge to make their science as much like physics as possible. It is, after all, the paradigm science (not necessarily the nicest science), even though biologists are often justifiably upset at the insensitivity of physics. At least, one can put matters this way. If we grant that there is nothing irreducibly teleological about thought in physics, then whether or not one thinks that biology should or should not be like physics, the very existence of a non-teleological science raises the question of whether any science, specifically biology, must be (or cannot help being) teleological?

So, how are we to set about answering these questions? What I want to suggest is that we might try using a source that the average philosopher of science considers anathema, certainly not something to give philosophical insight on contemporary science. I refer to the history of science, more particularly, in our case, I refer to the history of biology. Most philosophers draw a sharp distinction between what they refer to as the context of discovery and the context of justification, and then like Mrs Beeton and the hare, they refuse to ask any questions about where the objects of their attention come from. We saw in the last essay, for instance, that Popper rather ignores the process of arriving at ideas, preferring as a philosopher to concentrate on the analysis of given, contemporary science. And indeed, he is on record as saying that "to me the idea of turning for enlightenment concerning the aims of science, and its possible progress, to sociology or to psychology (or . . . to the history of science) is surprising and disappointing" (Popper, 1970, p. 57; see also Hempel, 1966). But no matter. Let us feel free to turn to biology's past to see if it throws light on biology's present. (Honesty compels me to admit that lesser philosophers have also accepted uncritically the distinction between the context of discovery and the context of justification, and the implication that history has no relevance to philosophy. The slight note of hysteria which enters my writings when I start proselytizing about the need for philosophers of science to turn to history is attributable to the same causes that made Paul so ardent a spokesman for Christianity after the journey to Damascus: uncomfortable memories of the past.)

4.2. ARTIFACTS AND ADAPTION

If we look at the history of biology, we see that 1859 was a watershed. That was the year of the publication of Charles Darwin's *On the Origin of Species*. Literally, biology was revolutionized (Ruse, 1979a). Before 1859 people believed that the organic world had been created miraculously by God. After 1859 people believed that the organic world was caused naturally by the action of unbroken law, and those who followed Darwin believed that the chief causal agent was natural selection working on random variation. From the point of view of teleology, 1859 and the publication of Darwin's *Origin* were crucial. And yet the *Origin's* importance is not straight-forward. I am reminded of Conan Doyle's marvelous Sherlock Holmes story about the theft of the race-horse Silver Blaze.

"Is there any point to which you would wish to draw my attention?"
  "To the curious incident of the dog in the night-time."
  "The dog did nothing in the night-time."
  "That was the curious incident," remarked Sherlock Holmes. (Doyle, 1902, p. 34.)

Affectionados will remember that this was Holmes's cryptic way of pointing out that the theft was an inside job, precisely because whoever was responsible was obviously a familiar figure. Similarly, the most important fact about the arrival of the *Origin* is that, from the point of view of the teleology in biology, *it did not make the slightest bit of difference*. Before Darwin, people cheerfully said that the eye existed in order to see. After Darwin, people cheerfully said that the eye existed in order to see. The causes were different perhaps, but the teleology was not!

"Is there any point to which you wish to draw my attention?"
  "To the curious way in which teleology changed after the *Origin*."
  "The teleology did not change after the *Origin*."
  "That was the curious way," remarked Sherlock Holmes.[1]

This all being so, let us ask first about the teleology of pre-*Origin* biology. Perhaps then, with this answered we can carry forward to modern biology. Now, the answer to questions about pre-*Origin* biology comes fairly rapidly. As every student of introductory philosophy knows well, people thought that they could speak teleologically about the organic world because they thought the organic world was teleological – it shows the ends put there by God for the benefit of humans and other organisms. (Whether everything ultimately reduced to human benefit was a nice point of natural theology.) Thus, as Paley (1802) was happy to argue, we can speak of the eye having the function of seeing, because the eye does have that function – it was designed (literally) that way by God. The eye, if you please, is one of God's *artifacts*.

I know no better method of introducing so large a subject, than that of comparing a single thing with a single thing: an eye, for example, with a telescope. As far as the examination of the instrument goes, there is precisely the same proof that the eye was made for vision, as there is that the telescope was made for assisting it. They are made upon the same principles; both being adjusted to the laws by which the transmission and refraction of rays of light are regulated. . . . For instance, these laws require, in order to produce the same effect, that the rays of light, in passing from water into the eye, should be refracted by a more convex surface than when it passes out of air into the eye. Accordingly we find that the eye of a fish, in that part of it called the crystalline lens, is much rounder than the eye of terrestrial animals. What plainer manifestation of design can there be than this difference? What could a mathematical instrument maker have done more to show his knowledge of his principle, his application of that knowledge, his suiting of his means to his end — I will not say to display the compass or excellence of his skill and art, for in these all comparison is indecorous, but to testify counsel, choice, consideration, purpose? (Paley, 1802, pp. 14–15.)

Moreover, let it not be thought that only professional natural theologians thought this way. In a discussion in 1834 of the adaptations of the kangaroo for feeding its young in Britian's leading science journal, the *Philosophical Transactions* of the Royal Society, we find Britain's leading zoologist, Richard Owen, calmly stated that such adaptations show "irrefragable evidence of creative forethought" (Owen, 1834).

Of course, in an important sense, pre-Darwinians were using a metaphor or analogy. Literally they thought the eye was one of God's artifacts; but they were modelling their understanding of it on human artifacts. We make the telescope, having our ends in mind. For this reason we can speak and think teleologically about the telescope. The eye, in many respects, seems very similar to the telescope. Therefore, it must have been made by God. "Things which are so design-like just don't happen by chance or blind law". So for this reason we can speak and think teleologically about the eye.

The key to pre-Darwinian biology, therefore, is the *artifact model* (Ruse, 1977). Because the organic world seems *as if* it were designed, it was felt permissible and appropriate to speak of it as actually being designed. And coming straight across the *Origin* to post-Darwinian biology, we see straight away that the teleological thought and language persists by virtue of the fact that the same point holds. Because the organic world seems *as if* it were designed, it is felt permissible and appropriate to speak of it as being designed. A molecule or the moon does not seem very much like an artifact, so we do not think such language appropriate in physics. The fins on the stegosaurus, however, are another matter. They look like turbo blades or the Heath Robinson contraptions that solar energy buffs are into. So why not talk and think that way?

Similarly, when working with optimality models, the biologist is faced with a situation or phenomenon which seems as if designed or thought up by someone like an engineer or economist who is concerned to get the maximum output for the minimum input, and so the biologist argues and reasons as if this really were the case. "In order to employ engineering optimization models the biologist tries to interpret living forms as in some sense the 'best'. In effect the biologist 'plays God': he redesigns the biological system, including as many of the relevant quantities as possible and then checks to see if his own optimal design is close to that observed in nature. If the two correspond, then nature can be regarded as reasonably well understood" (Oster and Wilson, 1978). Holling (1964) argues simply as if the foreleg and prey size of the praying mantis were a problem in engineering, for no other reason than because they seem as if they are such a problem. The world seems teleological; hence, teleological thought seems appropriate. For Charles Darwin himself incidentally, the continued use of teleological language came very naturally, because he had been brought up on Paley's *Natural Theology* (Darwin, 1969).

None of this is to deny| that pre- and post-*Origin* there is a difference. Pre-*Origin* the eye was caused by God. Post-*Origin* the eye was caused by natural selection. The teleology, however, remains the same. Lewontin, with his customary clarity, makes the point quite explicitly about modern teleology.

Much of evolutionary biology is the working out of an adaptationist program. Evolutionary biologists assume that each aspect of an organism's morphology, physiology and behaviour has been molded by natural selection as a solution to a problem posed by the environment. The role of the evolutionary biologist is then to construct a plausible argument about how each part functions as an adaptive device. For example, functional anatomists study the structure of animal limbs and analyze their motions by time-lapse photography, comparing the action and the structure of the locomotor apparatus in different animals. Their interest is not, however, merely descriptive. Their work is informed by the adaptationist program, and their aim is to explain particular anatomical features by showing that they are well suited to the function they perform. Evolutionary ethologists and sociobiologists carry the adaptationist program into the realm of animal behaviour, providing an adaptive explanation for differences among species in courting pattern, group size, aggressiveness, feeding behaviour and so on. In each case they assume, like the functional anatomist, that the behaviour is adaptive and that the goal of their analysis is to reveal the particular adaptation. (Lewontin, 1978, pp. 216–17.)

Let me sum up now what I have tried to say or hint at so far. I argue that the teleology in modern biology is analogical. The organic world seems as if it is designed; therefore we treat it as designed. The *artifact model* is the key to biological teleology. The phenomena of physics and chemistry do not seem as if designed. Therefore, we do not think teleological thought and language appropriate in those sciences. Of course, we do not today think biological phenomena really are designed – at least, not by the direct intervening agency of a miracle-working God. Rather, we think that natural selection working on random variation is the causal key. Therefore, if one wants to cash out a functional or teleological statement, "The *function* of the fins on the stegosaurus is to effect heat regulation", "The fins of the stegosaurus exist *in order that* its heat might be efficiently regulated", "The fins *solve the problem* (*serve the end*) of heat regulation", one must cash it out in terms of natural selection. The stegosaurus has fins because those of its ancestors which had fins survived and reproduced, and those that did not, did not. In short, the fins are an *adaption*, and to talk in functional language in biology is to refer to an *adaption*. If x has the function of y, then x is an adaptation, and y is adaptive — it helps survival and reproduction. Similarly, the foreleg of the praying mantis ultimately all comes down to a matter of survival and reproduction. Insects which went after prey of diameters significantly different from Tsin (β–α) simply did not pass on their genes.

This is the crux of what I want to say about biological teleology, including all of today's talk among biologists about optimality models. In a way it is all rather simple. But that does not mean it is not true. It makes sense of biology and of its history. Moreover, for once as a philosopher, I am not prescribing what biologists ought to do. Rather, I am describing what they do do. For instance, in one of the most highly praised books of the past two decades, the evolutionist G. C. Williams commits himself to precisely the position on teleology I have just unpacked and endorsed.

Whenever I believe that an effect is produced as the function of an adaptation perfected by natural selection to serve that function, I will use terms appropriate to human artifice and conscious design. The designation of something as the *means* or *mechanism* for a certain *goal* or *function* or *purpose* will imply that the machinery involved was fashioned by selection for the goal attributed to it. When I do not believe that such a relationship exists I will avoid such terms and use words appropriate to fortuitous relationships such as *cause* and *effect*.

Thus I would say that reproduction and dispersals are the goals or functions of purposes of apples and that the apple is a means or mechanism by which such goals are realized by apple trees. By contrast, the apple's contributions to Newtonian inspiration and the economy of Kalamazoo County are merely fortuitous effects and of no biological interest. (Williams, 1966, p. 9, his italics.)

Enough by way of defence. Let me now conclude with four corollaries which follow from my main theorem.

4.3. CONSEQUENCES AND AMPLIFICATIONS

First, one might ask why teleology is possible in biology? Why can a biologist profitably use a optimality model? In one sense I have already answered this question; in another sense I do not have to answer it. My answer is that teleology is possible because the organic world is design-like, and this is why optimality models work. My answer does not depend on a correct answer to the further question: Why is the organic world design-like, or perhaps, why does natural selection make the organic world design-like? But having exempted myself from the responsibility of having to give a correct answer (!), let me make two obvious suggestions, one which I confess I find rather less plausible than the other. The first suggested answer to the organic world's design-like nature rests on the putative fact that human design-processes and natural selection work in the same way — trial and error, over and over again. Think of the evolution (significant word!) of the automobile motor, a human artifact if anything is: If something worked, it was kept; if it did not, it was scrapped and something else tried. The principle is the same in organic evolution, even though this latter is much slower than the evolution of human artifacts. But the overall driving forces are identical. Clearly this suggestion is similar to Popper's theory of scientific theory growth (Popper, 1962, 1972). Indeed, it is basically Popper's position writ large, and I believe it suffers from the same difficulties which I brought up in the last essay. In particular, although human artifacts do obviously tend to have a trial and error history, it seems to me that human inventiveness cuts through much of the trial and error nature of natural selection. Selection really does work on random mutations. Human inventions are rarely built on random ideas. The disanalogy is crucial.

Possibly one might argue that the very act of creativity requires some sort of trial and error process; maybe one which is unconscious. I do not know, although it strikes me as a little implausible (Campbell, 1974). However, rather than persuing this line of thought here, let me turn abruptly to the second suggestion for explaining the design-like nature of the world. This is a suggestion to be found in David Hume's *Dialogues Concerning Natural Religion*, namely that irrespective of whether there is any similarity in the production of human artifacts and organic characteristics, the latter have to be design-like because if they were not, they simply would not work. The eye is like a telescope or some other human artifact for seeing, because if it were not like a telescope or related artifact, one would not see at all.

But where-ever matter is so poized, arranged, and adjusted as to continue in perpetual motion, and yet preserve a constancy in the forms, its situation must, of necessity, have all the same appearance of art and contrivance, which we observe at present. All the parts of each form must have a relation to each other, and to the whole: and the whole itself must have a relation to the other parts of the universe; to the element, in which the form subsists; to the materials, with which it repairs its waste and decay; and to every other form, which is hostile or friendly. A defect in any of these particulars destroys the form; and the matter, of which it is composed, is again set loose, and is thrown into irregular motions and fermentations, till it unite itself to some other regular form. If not such form be prepared to receive it, and if there be a great quantity of this corrupted matter in the universe, the universe itself is entirely disordered. (Hume, 1779, pp. 156–7.)

What Hume is saying here, and what I am endorsing, is the claim that, in order to have things working at all, certain 'problems' must be 'solved'. The world, specifically the organic world, had to solve these problems as a condition of its existence. Necessarily, therefore, it is design-like, because the problems are the same as, or akin to, the very sorts of things that humans want to solve. And this, I believe, is why biologists can use an artifact model. The world did not have to be design-like — large parts of it are not — but if we are to be around to worry about it, large parts of it had to be as if designed!

The second question which arises from my general position on biological teleology is one which asks about how hard-line biological teleology really is. Teleological explanation or understanding is, as we have seen, something which makes reference to ends. Teleologically, we understand the heart in terms of the end of pumping the blood; the stegosaurus's fins in terms of the end of heat regulation; and the praying mantis's forelimbs in terms of the end of prey capture. But this then raises the question always raised about teleological explanation: Are we trying to explain the past in terms of the future? Are we explaining the past or present forelimbs of the praying mantis in terms of the future prey capture? And if indeed we are, how then do biologists avoid the classic problems of teleology like that of the missing goal-object? (If one explains x in terms of future y, that is fine if y occurs; but what happens if x occurs, and then y does not?) In short, in the terminology I introduced just above, is biological teleology — especially when it comes to explanation and understanding — really 'hard-line', or is it all a little bit pseudo? (See Scheffler (1963) and Mackie (1966), for a full discussion of the problems of teleology.)

My own hunch, which I shall more state than argue for here, is that the teleology is fairly hard-line. I would argue that when the biologists says x exists in order to y, or the function of x is y or (y-ing), then the earlier x is indeed being explained in terms of the later y. Consider (for support) the following two passages by G. C. Simpson, a man whom we have recognized as one of the world's leading evolutionists. He writes:

In order to realize the new functions of a changed environment, an organism must, at the moment when the change or the occupation of a new environment begins, have at least some functions prospective with regard to the new environmental functions. This is merely a more technical way of saying that organisms can continue to live only under conditions to which they are already at least minimally adapted. (Simpson, 1953, p. 189.)

If a mutation, whether adaptive, non-adaptive, or inadaptive with respect to the adaptation of ancestral population, does become fixed and spread by selection it is adaptive from the start with respect to the descending populations. (Simpson, 1953, pp. 194–5.)

This, it seems to me, makes it all clear that x is being explained in terms of the later (future) y.

But what then about the missing goal-object problem and related problems? I think these do not really arise (or only very rarely arise) because, by the time that the biologist gets around to working out the function of x's, the y's (which are future to the x's) are now past too — in short, the biologist knows that they occurred and were not missing! Assuming that the stegosaurus explanation is correct, one can as a paleontologist explain the plates in terms of (later) heat regulation, because the heat regulation is in our past also, and cannot, therefore, go missing. As far as today's organic world is concerned, the biologist is still working on organisms essentially as they have gone, or are going past. The praying mantis of today is much like the praying mantis of yesterday. If the biologist wants to state categorically that at this moment the function of x is to do y and y really is still future, then frankly I think he/she is taking a gamble. She/he might come up against the missing goal object problem. But of course, practically speaking, the biologist is on safe ground. The future, at least for this year's generation, is usually like the past — eyes go on seeing, and legs go on walking — and so by the time that anyone gets around to deciding that the future is not like the past, the future is not the future any more. In other words, there is certainly nothing particularly fool-hardy about the biologist's thought and behaviour at this point. No one is caught up in beliefs about little men in the future pulling strings which affect the present. Biology is just about as cautious as everything else.

A third question raised by my analysis of teleology, concerns the matter of whether indeed all of the teleology of biology has been located and analysed. I would suggest that, in fact, it probably has not, and that in trying to find and discuss this extra teleology we can, incidentally, throw some light on a concept that keeps recurring in the philosophical literature on teleology, namely that of a *goal-directed system* (Sommerhoff, 1950; Nagel, 1961; Boorse, 1976b). The kind of paradigm which is taken in illustration of goal directedness is that of a torpedo which is programmed to strike a moving object, redirecting itself as the object changes place (Figure 4.4). I suspect that most of us who have worked on teleology, have a nagging feeling that goal directedness has, or ought to have, some sort of relevance somewhere. Some philosophers, indeed, have gone so far as to argue that all of the teleology of biology is analysable in terms of goal-directedness, meaning the

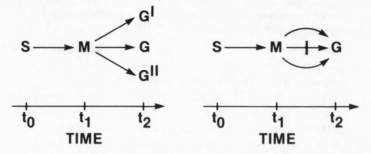

Fig. 4.4. Schematic representation of simple goal-directed systems. Suppose that an object X leaves the starting point S, directed towards G. However suppose also that upon arriving at the mid-point M the goal has moved to G^I or G^{II}, or alternatively that a barrier now blocks the direct passage from M to G. A goal-directed system is one which can take note of this new information, so that X gets redirected, towards the new place of the goal or around obstacles (as appropriate). In real life obviously goal-directed systems are usually very much more complex than these.

ability to get back on target despite disruptions. However, such a sweeping reliance on goal-directedness is clearly unwise (Ruse, 1971, 1973a). The assumption that the plates of the stegosaurus exist in order to effect heat-regulation, says nothing about how the stegosaurus would react were something untoward to happen. Similarly, if a praying mantis were faced with nothing but prey not of the optimum size, it might be unable to react and thus starve to death. But I think that goal-directedness is nevertheless important in biological teleology, and my central analysis shows how.

Consider: We get the teleology of functions in biology by analogy from human teleology, namely the teleology of human artifacts. Is there any other kind of human teleology? Clearly there is, namely our own teleology. We ourselves strive to achieve certain ends, just as we try to impregnate ends into our artifacts. I aim, teleologically, to write a book, just as the book's dust jacket serves the end of keeping the book reasonably clean. I see no reason why we should not, and indeed would rather expect that we would, read this other kind of teleology — the teleology of the individual, intending being — into the organic world also. But how would we recognize this teleology, or rather what would lead us to impute such teleology to the non-human world? I suppose the existence of brains in other organisms would go part way, but I suspect that identifying goal-directed behaviour is also a very important, perhaps crucially necessary, factor. Suppose I am teleologically oriented towards something, say getting a PhD. The real mark that I have that PhD as a goal is the fact that I strive for it, despite all obstacles like boring professors, irrelevant language exams, inability to get up in the morning, and so forth (more shades of the Ruse Past!). Similarly, we see a teleology in the organic world, over and above the teleology of functions, when the wolf-pack shows a flexible strategy to bring down the much larger moose, and when the insect colony regroups and moves in the fact of attack, and so forth. In other words, by analogy we see goal-directed behaviour as teleological, of a kind different from that of function. There are, therefore, two kinds of biological teleology: the teleology of functioning artifacts and the teleology of intending beings, where goal directedness is taken as an important mark of such intention.

I am not sure, however, that we see all goal-directed behaviour as teleological, and this is why supposed problem examples of goal-directedness, drawn from the inorganic world, do not worry me. As we get farther from human intelligence and its ability to conceive of and act towards goals, we see less teleology (of this second kind). Take something like sweating and shivering, which is generally conceded to fill the conditions for a goal-directed system. Is this teleological? A body capable of sweating and shivering is certainly teleological inasmuch as such processes are adaptations (which they are!). But is there more teleology, namely the kind I find in the coordinating wolf pack? I rather doubt it — no one thinks that any intelligence is involved in keeping the body at constant temperature through sweating and shivering. I suspect that there is a grey area about where this second form of teleology operates. We find it in dogs. We do not find it in trees. Do we find it in earth worms? In part, this is an empirical question. How goal-directed are earth worms capable of being — particularly, how goal-directed in terms of *behaviour* are they capable of being? After that, the imputation of such teleology is all a bit arbitrary. But then, of course, this element of arbitrariness is characteristic of metaphor or analogy. Some people are prepared to push metaphors further than others. A grey area where there are disagreements confirms my position, rather than refutes it.

Fourth and finally, let me ask the following questions about my analysis of biological teleology, ones which raise matters of reduction and eliminability. Could we get rid of the teleology of biology? Could biologists function efficiently without optimality models? And even if we could achieve such a 'purified' biology, would we want it or use it? My own suspicion is that a non-teleological biology is possible, but not really that exciting or desirable, although at least with respect to optimality models, I put myself here in opposition to some important biologists (see Allen *et al.*, 1977). We could certainly talk always in terms of past causes (efficient causes) and drop all talk of function and so forth. But we could not then say as much. We might, for example, talk about the embryological development of the heart, talking in terms of cell-division and so forth, but we could not ask what the heart was for. Similarly, we might look at the stegosaurus in terms of development or perhaps even in terms of past gene ratios and the way they altered, but we could never get round to asking what the plates on the back are for. And this seems to be an interesting question. As also are questions which we can formulate and answer using optimality models. Why do praying mantises always attack prey of a certain size? Why do the hymenoptera have certain caste features? And so forth. If nothing else, if we drop these models, we lose our incredibly powerful heuristic guide.

Moreover, if one were determined to excise the peculiar nature of biology, simply to switch to talk of 'adaptation' would be no solution, because the forward-looking teleological aspect of biology still then remains. The whole way of looking at organisms in terms of design and artifacts has to be dropped. But why would one want to do this? Biology is not physics, nor should it have to be. If there were no good reason for biology's being different, then one would expect them to be the same. This is why I feel justified in neglecting neo-vitalistic attempts to separate biology and physics on such pseudo-grounds as the purportedly peculiarly complex nature of biology. (See, for instance, Glass, 1963; Polanyi, 1968.) In pertinent respects, biology is no more complex, or unique, or what have you, than physics. Hence, in my first essay I felt no compunction about pushing ahead and arguing that, in the important respects I detailed, the synthetic theory of evolution is very much like a theory of physics and chemistry. Convincing reasons to the contrary, I expect biology to be like physics. Biologists do not have peculiar ways of thinking simply because they are biologists.

But when we come to teleology, I believe that we do have good reason for separating physics and biology, as they stand now. And the reason for this is that biology refers to phenomena that are design-like and physics does not, and biology as a science tries to capture and utilize this distinctively teleological facet of the biological world in descriptions, explanations, optimality models (whether heuristic or explanatory) and the like. Moreover, interestingly, at the borderline, if anything, it is physics that gives. Where biology and physics meet, we do not get a total rejection of teleology. Rather, conversely teleology seeps in. At least, we find molecular biologists talking about such things as the need "to crack the genetic code" (Watson, 1975). And, if that is not to talk in terms of design, I do not know what is. In short, the teleology of biology is here to stay. Given its power in the hands of evolutionist, for instance in finding out the purpose of the plates on the back of the stego-saurus, or in understanding the parts of the praying mantis, let us be thankful for this fact.[2]

NOTES

[1] Having so strongly committed myself to the importance of history, honesty forces me to admit that it is highly unlikely that Holmes ever actually engaged in this precise exchange. His knowledge of and interest in science were selective, to say the least. Apparently, until Dr Watson enlightened him, he was ignorant of the Copernican Revolution.

[2] The philosophical reader will recognize that in respects, my stance on biological teleology supports the general thesis on the use of analogy or metaphor propounded by Max Black (1962), namely that in some sense through the use of metaphor (in biology, what I have called the 'artifact model') we get a kind of understanding impossible without the metaphor (see also Ruse, 1977a).

THE MOLECULAR REVOLUTION IN GENETICS

"We wish to suggest a structure for the salt of deoxyribose nucleic acid (DNA). This structure has novel features which are of considerable biological interest" (Watson and Crick, 1953, p. 737). This brief paragraph begins what is probably the most famous biological paper of this century; what is indeed possibly the most famous paper since those of Mendel in the 1860s or even the Darwin–Wallace papers announcing natural selection, read before the Linnean Society in 1858. But why should a short contribution by two junior researchers, James Watson and Francis Crick, about the structure of a molecule, a physico-chemical entity, be an important paper in *biology*? Quite simply because the DNA molecule is found in organic cells precisely where one would expect the units of heredity to be, and because their model of the molecule strongly suggested that they had hit upon and understood the key to the molecular basis of heredity. As Watson and Crick laconically remarked at the conclusion of their paper: "It has not escaped our notice that the specific pairing we have postulated immediately suggests a possible copying mechanism for the genetic material" (ibid.). Thus started the molecular revolution in genetics, or perhaps one might more fairly acknowledge those upon whose shoulders Watson and Crick stood, and say that thus the molecular revolution in genetics came of age (Olby, 1974; Judson, 1979).

Jumping to 1980 and looking at the present state of the science, we see that the molecular foundations of the principles of heredity are almost breath-takingly simple in their essential outlines. Those philosophers like William Whewell who have argued that simplicity is a cardinal virtue of the greatest of scientific theories would really feel vindicated by molecular genetics (Whewell, 1840). Following Watson and Crick, it is agreed that the DNA molecule is indeed the carrier of the information of heredity and is, thus, the ultimate biological unit of function (Watson, 1975). The molecule, which occurs in the nuclei of cells, is a polymer of deoxyribose sugars (i.e., chain of such sugars) joined by phosphate links. Attached to each sugar, as a side chain, is a nitrogen-containing base, which must be one of four kinds: adenine (A) or guanine (G) (purines), or thymine (T) or cytosine (C) (pyrimidines). These bases, together with the sugars and phosphates, are called 'nucleotides'. The DNA molecule does not occur on its own; rather it is always coiled around a mate in a 'double helix'. The bases on complementary molecules in the helix are paired: adenine always pairs with thymine, and guanine with cytosine. Potentially, the nucelotides in a DNA molecule can occur in any order whatsoever and, thus, a DNA molecule can carry as much information as one wants. It is exactly the same simple principle at work as that which we ourselves use when we alter around and repeat letters of the alphabet in order to write books on every and any topic under the sun. (See Figure 5.1). All of this information can be found in any textbook on genetics, for instance, Ayala and Kiger, 1980. Readers in search of a suitable introduction might consult Mertens and Polk, 1980.)

The process of *replication*, the reproducing of the DNA molecule itself (necessary when new cells are formed), is totally straightfoward. The original double helix 'unzips', and as it does so the two strands attract free nucleotides in the cell, thus acting as templates to make complementary chains. One therefore gets two identical double helices. The way in which the DNA molecule acts as the unit of function is hardly more complex. Essentially, organisms are made up of proteins, which can be either structural proteins (the 'building blocks'), these going to make up the walls of the cells and so forth, or enzymes, which are catalysts, these causing the cells to go about their chemical activities of breaking down and building up in an orderly manner. Proteins are all 'polypeptide' chains or combinations of such chains. The numbers of the links in a polypeptide chain can run to the hundreds, each link is an amino acid, and the nature, order, and number of such acids determines the particular chain and, thus, the particular protein. DNA makes protein *via* an intermediary macromolecule, riboneucleic acid (RNA), which differs from DNA only in having uracil (U) wherever thymine occurs in DNA. First in the manufacture of protein, there is the process of *transcription*, whereby RNA is formed from the DNA template and, thus, picks up the information of the DNA. Then this 'messenger' RNA goes to other parts of the cell, the ribosomes, where the process of *translation* occurs. With the aid of more RNA, amino acids in the cell are picked up, brought over to the ribosomes, and linked together in the way dictated by the messenger RNA. Thus, the information of the DNA is used to make distinctive proteins (Figure 5.2).

Completing the story, we should note that there are twenty amino acids. Since DNA and RNA have only four bases, one has to have a 'code' whereby a number of bases in specific sequence (a 'codon') corresponds to any particular amino acid. If this were not so, then the messenger RNA could not specify exactly which acid is needed in exactly which order. In order to code

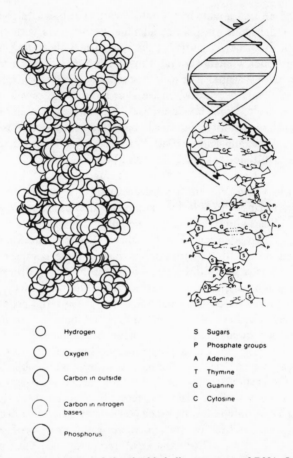

Hydrogen		S	Sugars
Oxygen		P	Phosphate groups
		A	Adenine
Carbon in outside		T	Thymine
		G	Guanine
Carbon in nitrogen bases		C	Cytosine
Phosphorus			

Fig. 5.1. "Two representations of the double-helix structure of DNA. Left: a 'space-filling' diagram. Right: additional details. The molecule consists of two complementary chains. The outward backbone of the DNA molecule is made of alternating deoxyribose sugar (S) and phosphate (P), chemically bound to one another. Nitrogen bases connected to the sugars project towards the centre of the molecule. There are four kinds of bases: adenine (A), cytosine (C), guanine (G), and thymine (T). Two chains are held together by hydrogen bonds between complementary bases, but A can only pair with T, and C only with G" (from Ayala and Valentine, 1979, p. 53).

for twenty amino acids (together with the information to stop translation), if one is given only four bases, then one needs a minimum of three bases in sequence for any amino acid. Codons are, in fact, triplets, and because 4^3

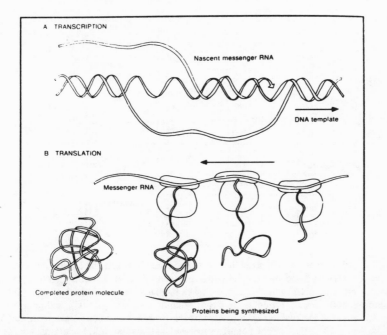

Fig. 5.2. "The process of transcription and translation. (A) Transcription. One strand of the DNA helix serves as a template for the synthesis of a complementary molecule of messenger RNA (RNA stands for 'ribonucleic acid' a molecule quite similar to a single chain of the DNA double helix). (B) Translation. The messenger RNA becomes attached to certain bodies, known as ribosomes, that mediate the synthesis of proteins. The sequence of codons in the messenger RNA determines the sequence of amino acids in the protein" (from Ayala and Valentine 1979, p. 56).

(the maximum possible number of triplets) is larger than 20 + 1 (the minimum number of different bits of information needed at the ribosome) there is duplication of codons. Molecular geneticists have now 'cracked' the genetic code completely (Figure 5.3).

In outline, this is all there is to molecular genetics although, of course, much important and exciting work has been done and remains to be done in filling out the details. Mutation, for instance, finds a ready explanation. Suppose one has some alteration of the information on the DNA molecule, for example a mistake in replication. This mistake could then reflect itself ultimately in a change in the proteins produced. And continuing with a kind of domino effect, if the organism could no longer produce a certain protein, say a certain sort of enzyme, then possibly certain other proteins, say

SECOND LETTER

	U	C	A	G	
U	UUU ⎫ Phe UUC ⎭ UUA ⎫ Leu UUG ⎭	UCU ⎫ UCC ⎪ Ser UCA ⎬ UCG ⎭	UAU ⎫ Tyr UAC ⎭ UAA ⎫ Stop UAG ⎭	UGU ⎫ Cys UGC ⎭ UGA Stop UGG Trp	U C A G
C	CUU ⎫ CUC ⎪ Leu CUA ⎬ CUG ⎭	CCU ⎫ CCC ⎪ Pro CCA ⎬ CCG ⎭	CAU ⎫ His CAC ⎭ CAA ⎫ Gln CAG ⎭	CGU ⎫ CGC ⎪ Arg CGA ⎬ CGG ⎭	U C A G
A	AUU ⎫ AUC ⎪ Ile AUA ⎬ AUG Met	ACU ⎫ ACC ⎪ Thr ACA ⎬ ACG ⎭	AAU ⎫ Asn AAC ⎭ AAA ⎫ Lys AAG ⎭	AGU ⎫ Ser AGC ⎭ AGA ⎫ Arg AGG ⎭	U C A G
G	GUU ⎫ GUC ⎪ Val GUA ⎬ GUG ⎭	GCU ⎫ GCC ⎪ Ala GCA ⎬ GCG ⎭	GAU ⎫ Asp GAC ⎭ GAA ⎫ Glu GAG ⎭	GGU ⎫ GGC ⎪ Gly GGA ⎬ GGG ⎭	U C A G

FIRST LETTER   THIRD LETTER

Fig. 5.3. "The genetic code gives the correspondence between the 64 possible codons in messenger RNA and the amino acids (or termination signals). The nitrogen base thymine does not exist in RNA, where uracil (U) takes its place; the other three nitrogen bases in messenger RNA are the same as in DNA; adenine (A), cytosine (C), and guanine (G). The 20 amino acids making up proteins are as follows: alanine (Ala), arginine (Arg), asparagine (Asn), aspartic acid (Asp), cysteine (Cys), glycine (Gly), glutamic acid (Glu), glutamine (Gln), histidine (His), isoleucine (Ile), leucine (Leu), lysine (Lys), methinoine (Met), phenylalanine (Phe), proline (Pro), serine (Ser), threonine (Thr), tyrosine (Tyr), tryptophane (Trp), and valine (Val)" (from Ayala and Valentine, 1979, p. 57).

structural proteins, could become unavailable, and this fact could reflect itself right the way up to the visible phenotype (Watson, 1975).

5.1. SCIENTIFIC ADVANCE: REDUCTION OR REPLACEMENT?

So much for scientific facts. The question I want to ask is, whether this molecular revolution in genetics has any interest and relevance to those of us who want to understand the ultimate nature and progress of science? And my answer is that I think it does — a most particular interest and relevance — especially since molecular genetics did not just arrive in virgin land, totally empty of any earlier denizens. As we know full well from the earlier essays in this collection, by the 1950s in biology there was already a full-developed theory dealing with the ultimate causes of organic characteristics and the way in which these characteristics can be transmitted from one generation to the next: Mendelian genetics. In other words, just as the first white people on the North American shore came face to face with the Indians, so the molecular geneticist came face to face with the Mendelians. But this means we have a situation which seems to bear directly on a hotly debated question about the nature of scientific progress or change. Let me explain.

It is generally agreed that sometimes in scientific revolutions, or even in fairly minor changes, which we would not normally label 'revolutionary', we get an older theory or position entirely pushed to one side by a newer theory or position. Thus, for instance, in the Copernican revolution all the old ways of thinking had to go: a geocentric, small universe; crystalline spheres and perfect heavens; circular motions and epicyles (Kuhn, 1957). Similarly in the Darwinian revolution, beliefs dear to the hearts of Special Creationists vanished (Ruse, 1979a). In these cases, the newcomers entirely vanquished the old. Let us speak of situations like these as involving *replacement*. Sometimes, however, the changes in science seem nothing like as Draconian. In fact, in a way the old-timers and their views do not get destroyed; rather they get assimilated into the new-comers and their views, – they are made part of a grander whole. Newton, for instance, did not throw out the insights of Kepler and Galileo. He brought them beneath his laws. Similarly, the kinetic theory of gases did not destroy older work, like Boyle's law. Somehow it incorporated Boyle's law into itself (Holton and Roller, 1958). Let us speak of this process as *reduction*.

Now, those of us who subscribe to what I referred to in the first essay as the 'received view' of scientific theories (often we are called 'logical empiricists') feel that we can give a ready and convincing theoretical analysis of scientific theory reduction (Nagel, 1961; Hempel, 1966; Yoshida, 1977). We see scientific theories as axiom systems, with all claims stemming from a few basic principles. We argue, therefore, that in a reduction (say involving an older theory T_1 and a newer theory T_2), what happens is that the older theory (specifically its axioms) is shown to be a *deductive* consequence of the newer theory. Thus the older theory is not lost or rejected, but shown to be a consequence of a greater whole: Kepler's and Galileo's laws can be derived from Newton's laws, and Boyle's law can be derived from the kinetic theory of gases. Speaking a little more carefully, we admit that sometimes a reduction is not from a newer theory alone; rather it is from the newer theory plus one or two other bits and pieces. Just as in a regular theory, one needs translation rules to go from talk of theoreticals to talk of observables, so in a reduction one similarly needs translation rules or 'reductive functions' to go from the language of one theory to the language of another theory – say, from talk of molecules to talk of heat.

Ernest Nagel puts matters this way:

Accordingly, when the laws of the secondary science do contain some term 'A' that is absent from the theoretical assumptions of the primary science, there are two necessary formal conditions for the reduction of the former to the latter: (1) Assumptions of some kind must be introduced which postulate suitable relations between whatever is signified by 'A' and traits represented by theoretical terms already present in the primary science . . .

(2) With the help of these additional assumptions, all the laws of the secondary science, including those containing the term 'A' must be logically derivable from the theoretical premises and their associated co-ordinating definitions in the primary discipline. (Nagel, 1961, pp. 353–4.)

Like many of the tenets of logical empiricism, this whole theory of scientific theory reduction has been subjected to strenuous attack in the past two decades. Whatever its formal adequacy, it has been argued that as a matter of historical record, it simply never fits the actual transitions from older positions to newer positions. Thus, Paul Feyerabend (1970) argues that, in fact, Kepler's and Galileo's laws cannot be derived from Newton's laws. For instance, if H is the height above ground of processes being described and R is the radius of the earth, then so long as the ratio H/R is finite, Galileo's laws cannot follow. And obviously H/R is always finite. Similarly, Thomas S. Kuhn argues that there was no reduction in another favourite example of the logical empiricists, that of Newtonian physics to Einsteinian physics. The concepts of the two sciences are fundamentally dissimilar. "Newtonian mass is conserved; Einsteinian is convertible with energy" (Kuhn, 1970, p. 102). Hence, argue these and similar thinkers, what we get in science is replacement, more replacement, and yet more replacement. The course of science is a little like the course of Elizabeth Taylor's love-life (Figure 5.4).

Logical empiricists are tenacious folk. In the face of all the criticism, a certain regrouping of the forces has been necessary. However, there is anything but headlong retreat. In particular, it has been argued, most vigorously by Nagel's student Kenneth Schaffner, that, with but a slight revision, the logical empiricist account of reduction can be retained (Schaffner, 1967, 1976). Perhaps one cannot always derive T_1, the older theory, from T_2, the newer theory, but one can derive $T_1{}^*$ from T_2, where $T_1{}^*$ is as it were a corrected version of T_1 — like T_1 only a little more accurate. (Schaffner speaks of 'strong analogy' existing between T_1 and $T_1{}^*$.) Thus, one might say that what one derives from the kinetic theory of gases is not Boyle's law but van der Waal's equation, where the latter is slightly more accurate than the former. But it is still reasonable to speak of 'reduction' because Boyle's law

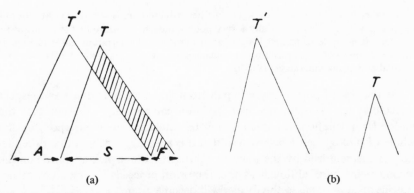

(a)   (b)

Fig. 5.4. (a) Here we have the relationship between two theories as envisioned by the logical empiricists (T is superceded by T'); S is material explained by both theories; A is material explained only bv T' and F is the false information predicted by T. (b) Here we have the Kuhn–Feyerabend vision (from Feyerabend, 1970, p. 220).

has hardly gone the way of Ptolemy or Special Creationism. We see that it holds roughly within certain limits, and the kinetic theory of gases explains why it holds as it does, and no further. The important thing about this weaker notion of reduction is that it still helps one to see the continuity between the old and the new, which simple talk of 'replacement' conceals or denies.

5.2. WHAT KIND OF REVOLUTION OCCURRED IN GENETICS?

In the light of this debate about reduction and replacement, the molecular revolution in genetics draws our attention. Surely something so fresh and new and apparently so important will guide the ways of philosophers? Conversely, dare one hope that the philosophical debate might throw some light on the exact nature of the molecular revolution? Is it as far-sweeping and significant as some of its protagonists have clearly thought and claimed, or is it somewhat limited in its importance for much of biology? Certainly some leading evolutionists have downplayed its significance. For instance, Ernst Mayr is on record as stating:

To be sure, most of the phenomena of functional biology can be dissected into physical-chemical components, but I am not aware of a single biological discovery that was due to the procedure of putting components at the lower level of integration together to achieve novel insight at a higher level of integration. No molecular biologist has ever found it particularly helpful to work with elementary particles.

> In other words, it is futile to argue whether reductionism is wrong or right. But this one can say, that it is heuristically a very poor approach. Contrary to the claims of its devotees, it rarely leads to new insights at higher levels of integration and is just about the worst conceivable approach to an understanding of complex systems. It is a vacuous method of explanation. (Mayr, 1969b, p. 128.)

At the risk of putting myself apart from those whom I admire and respect incredibly, let me state my own position categorically. I believe that the molecular revolution in genetics was of fundamental importance for the whole of biology and I believe also that it is a paradigm case of theory reduction as characterized by the logical empiricists. Let me pick up a little on the second point here, although as the discussion proceeds I think some of my reasons of subscribing to the former will become clearer.

The crucial issue, obviously, is whether we can get the kinds of claims that the Mendelian geneticists want to make about their objects of interest, Mendelian genotypes and phenotypes, from the kinds of claims that the molecular geneticists want to make about their objects of interest, DNA molecules, polypeptide chains, and so forth. Granting that everything is rather loose, a matter to which I shall be returning later, and granting also what has been acknowledged above, namely that one will need appropriate translation rules, claims made in the Mendelian realm seem to be derivable from claims made in the molecular realm. Clearly, before one can do anything else, one needs some sort of rule identifying the molecular equivalent of the Mendelian gene — one can hardly have Mendel's first law without it! But equally clearly, the identification is going to involve a DNA molecule or a piece of one and, indeed, what we find is that Mendelian genes can be and are identified with functioning segments of DNA molecules. "A gene ... is a functional unit; it usually corresponds to a section of the DNA chain coding for an amino acid sequence in a protein" (Dobzhansky, 1970, p. 33). In fact, so readily is this identification made that molecular geneticists speak quite calmly and readily of molecular 'genes', meaning segments of DNA (Watson, 1975). With this identification, something like Mendel's first law comes quite naturally, because in their realm molecular geneticists are no less committed to its essential truth than are Mendelian geneticists in the purely biological realm. Molecular geneticists think that DNA molecules get passed on in much the way that Mendelian geneticists think their units are passed on: each offspring gets one of any of its functioning DNA strips from one parent (such strips being produced by replication and thus identical to parental strips) and each offspring gets the complementary strip from the other parent; moreover, there is the randomness in transmission which we find in the biological realm

(Ayala and Kiger, 1980). Hence, as I claim, with the appropriate dictionary (Functioning strip of DNA — Mendelian gene), we get the biological claims following naturally and deductively from the molecular claims.[1]

Similarly, if we are dealing with some particular Mendelian claim, say that a certain gene causes blue-eye colour or whatever, molecular and biological levels can be linked smoothly and without distortion, just so long as one has (what one expects to have) appropriate ways of linking molecular cell products, certain proteins for instance, with gross phenotypic effects. Concentrating on the general analysis, we can lay matters out as follow: If we let ma and Mb stand for 'a is a molecular gene' and 'b is a Mendelian gene', respectively, then our connection of identity is $ma = Mb$. Suppose now that we have a molecular claim that '$ma \to pc$' where 'pc' stands for 'c is some kind of molecular phenomenon' and '$\to$' represents some kind of causal connection. In other words, suppose we have a molecular claim that a certain gene has a certain causal effect, say producing a particular protein. If we can now either identify the molecular phenomenon with the phenotype, or more likely show that the phenomenon (i.e., the protein) leads to some sort of phenotypic effect, Pd, then since we now have '$pc \to Pd$', by assuming a number of obvious laws about causality, and so forth, we can derive deductively a Mendelian claim that "$Mb \to Pd$", that is that a certain Mendelian gene leads to a certain phenotypic effect.

Let us also note that there seems every reason to think that Mendelian relationships, for instance 'dominance' and 'recessiveness', have molecular equivalents and that, consequently, any Mendelian claims about such relationships can be derived from molecular genetics. One functioning strip of DNA can quite blank out the effects of another strip which, after all, is what dominance/recessiveness is all about. (See Schaffner (1976) for a detailed spelling out and justification of this claim.) Hence, what I would conclude is that at the molecular/Mendelian interface we seem to have a paradigmatic example of one theory being reduced, or reducible, to another. Both general and more particular claims of Mendelian genetics can be deduced from molecular genetics, and there seems no barrier to a like deduction of any claim of this biological theory of heredity.

Moreover, I do want to emphasize that this is all a great deal more than just a set of a priori flights of fancy, or of the kinds of promissory notes so beloved by philosophers about what 'could be done in principle'. Some parts of the reduction/deduction have been worked through in great detail in specific instances. Take the case of sickle-cell anaemia, and the Mendelian biological claim that the heterozygote has increased resistance to malaria.

Obviously what is needed, and what has already been proposed, is an identification of the Mendelian genes involved with molecular genes, in particular with those responsible (in part) for the production of the haemoglobin, the molecule which carries the oxygen in the blood. In fact, haemoglobin, which consists essentially of two pairs of polypeptide chains (α and β), is thought to be the product of two (non-allelic) genes (Figure 5.5). The sickle-cell mutant

Fig. 5.5. Schematic representation of the molecular structure of human haemoglobin. The tetramer structure (α_1, α_2, β_1, β_2) is depicted. Each polypeptide chain sub-unit has an associated haem group, indicated by the disk buried within the substance of the subunit. Oxygen binds to the ferrous iron of the haem group. The secondary structure of the amino acid sequence provides rigid helical regions and bends (Fraser and Mayo, 1975, p. 104).

(i.e., molecular gene mutant) causes the substitution of one amino acid in the β-chain for another amino acid (valine for glutamic acid). This substitution leads to 'stacking' of the molecules into uniform alignment, which increases the viscosity of the haemoglobin, which leads in turn to the characteristic sickle-shaped distortion of the cell. (See Figure 1.2.) In the heterozygote, because of the existence of some normal haemoglobin, one gets less stacking in a cell, which can thus function, but one still gets some increased viscosity. Parasitical infection by the falciparum malarial organisms of such a heterozygote cell increases sickling because the parasite uses oxygen, thus further increasing viscosity. It is then believed that the infected, much-sickled cell is removed by the body by phagocytosis. Thus, the malarial parasite sows the seeds of its own destruction. And obviously here we are out at the end of the causal chain with a biological claim about the increased resistance to malaria of the heterozygote! If this is not a specific example supporting the general thesis about reduction, I do not know what would be (Figure 5.6).

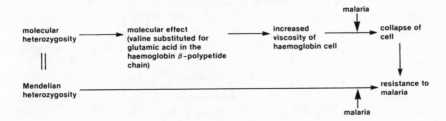

Fig. 5.6. Schematic representation of the relationship between molecular and biological levels in the case of sickle-cell anaemia. Here one is identifying the molecular hetero-zygosity with the Mendelian heterozygosity (mh = Mh) and suggesting that the collapse of an infected cell is either equivalent to or the cause of malarial resistance (in my diagram, cc → mr). Given the other proven causal connections (mh → me; me → iv; iv + mal → cc), we can infer the biological claim that heterozygosity leads to malarial resistance (Mh → mr). I am not absolutely sure that one wants to say that all of the top line is molecular (i.e., physico-chemical). Perhaps increased viscosity of a sickled cell gets one to the biological realm again. If this is the case, the proper biological claim is Mh → iv; but obviously this modification does not affect the claims about deduction/reduction.

5.3. BUT DID 'STRONG' REDUCTION REALLY OCCUR?

And yet, there is a nasty fly in the ointment — a molecular fly in the bio-logical ointment. There is good reason to think that Mendelian genetics is not quite so easily and neatly reducible from molecular genetics as I have implied. Why is this? Remember back to the first essay, where I introduced Mendelian genetics. The thing which really distinguished it from other theories of hered-ity, like Darwinian pangenesis for instance, is that it is *particulate* (Dunn, 1965; Sturtevant, 1966; Provine, 1971). One cannot chop up the Mendelian gene. It may change, but barring this, it is passed on intact from generation to generation. In particular, the phenomenon of crossing over, which leads to the shuffling in each generation of gene combinations on the same chromo-some, supposedly occurs only between genes and not within them. However, in molecular genetics there is no reason at all why the DNA molecule should not be broken right in the middle of a functioning unit, perhaps because of some such phenomenon as crossing over. The bonds between nucleotides within a functioning unit are on par with bonds between units. In other words, in an important sense the molecular gene, the functioning DNA segment, is not particulate. But this surely means that one can hardly hope for a full-blooded reduction in the logical empiricist sense, that is deduction of one theory from the other. If molecular genetics and Mendelian genetics make conflicting claims, then one cannot expect to find that the one is the logical consequence of the other. Specifically, one cannot deduce a theory with a particulate gene concept from a theory with a non-particulate gene concept.

What is one to do? Apparently, two options present themselves. On the one hand, one could throw up one's hands in despair, fearing that a reduction will never be possible and that the attempt to force one could only lead one into 'revisions' of biology that are totally removed from anything resembling real science. On the other hand, one could oneself attempt such a reworking of biology — particularly of Mendelian genetics — to see if something could be derived from molecular genetics which was, nevertheless, in important respects, still like Mendelian genetics. In this case one would have an artificially created 'corrected' biological theory with strong analogy to Mendelian genetics. One could therefore argue that one has the weaker form of reduction that logical empiricists (like Schaffner) have been proposing in recent years. One may not have 'strong' reduction (i.e., deduction); but one does have 'weak' reduction.

I am sure that a number of readers will be drawn to the first alternative. The second option seems to steer dreadfully close to that all-too-common kind of philosophizing which takes as a paradigm an idealized statement that no scientist would ever dream of holding, like "All swans are white", and then proceeds to build thesis and counter-thesis galore upon it. In a moment I will look at the articulate views of a philosopher, who has this very feeling about attempts to find some sort of reduction in biology. However, let me first reveal that the thinker, who does indeed seek a reduction relationship in genetics, is not necessarily trapped into taking either of these rather unattractive options or alternatives. Looking at the history of biology in this century we see that the biological geneticists themselves did not sit still. They found the classical Mendelian gene concept inadequate, *for biological reasons* (Carlson 1966). Hence, they themselves were lead to make revisions to their theory, and in the light of these revisions I would suggest that talk of 'reduction' in the standard logical empiricist sense (i.e., 'strong' reduction) once again become appropriate and illuminating — and no 'corrections' simply for the sake of philosophy are necessary at all.

Briefly covering the relevant biology, as far back as 1925 it began to be recognized that the classical gene concept, treating genes as particulate rather like beads on a string, failed to account adequately for all the niceties of biological inheritance. In particular, Sturtevant (1925) reported on the 'position effect', namely that the position and order of genes on a chromosome is pertinent to their effect on the phenotype. Two identical alleles next to each other (on the same chromosome), heterozygous to other alleles, can have a stronger effect than the same alleles homozygous to each other on complementary chromosomes (Figure 5.7). Problems like this, inexplicable

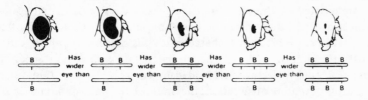

Fig. 5.7. This figure compares the eyes of female flies (*Drosophila*) with different numbers and arrangements of what is presumed to be the same gene, namely that which occurs at the so-called 'Bar locus'. The third and fourth from the left show the position effect. We have the same genes in different arrangements, and get different effects (i.e., different numbers of facets in the eye). If the Mendelian genes were totally particulate, like beans in a bag, one would not expect this effect. (Picture from Rothwell, 1979, p. 288.)

given the concepts and principles of classical Mendelian genetics, continued to mount through the years, until in the 1950s, for biological reasons, the whole particulate Mendelian gene concept fell apart and biological geneticists were lead to revise their entire position in crucial respects. In the place of classical Mendelian genetics, one now had 'fine structure genetics' or 'transmission genetics'.

What led to the overhaul? Simply, that geneticists started to use more sophisticated techniques and turned increasingly to organisms peculiarly suited to the examination of the most intimate aspects of the biological unit of inheritance. Obviously, if one is to infer the nature of the gene by studying the effects of breeding experiments – and this, after all, is just what geneticists do – then one is better off studying Drosophila than elephants. Fruitflies breed rapidly, and so one can readily test out hypotheses about crossing over, mutation, dominance, and the like. Also, one can more easily keep environmental conditions stable for fruitflies than one can for elephants. But even fruitflies have limitations. Therefore, a number of workers, particularly Seymour Benzer, turned to yet-more-rapidly reproducing organisms, like bacteria (e.g., *Escherichia coli*) and the viruses (often called 'phages') which attack these bacteria. Apart from speed, which enables one to pick out rare and interesting phenomena quickly, such organisms have other virtues, like being haploid (i.e., only half sets of chromosomes) and so one can avoid such problems as those that are caused by dominant genes masking other genes (Carlson, 1966; Whitehouse, 1965).

The import of all of this is that, using purely biological techniques, Benzer was able to show that the biological unit of function — what had hitherto been represented by the old Mendelian gene — does, in fact, on occasion get broken apart. In other words, it is no longer reasonable to claim that the ultimate cause of biological characteristics is something which can never be cut into. Biological genetics has to modify its particulate stance. Indeed, Benzer suggested replacing the Mendelian gene concept with three concepts: the 'cistron', the unit of function; the 'muton', the unit of mutation; and the 'recon', the smallest unit of crossing over (i.e., that which really cannot be split up). As a matter of fact, generally biologists have not really bothered too much with the terminology, but the findings are acknowledged by all. And what is absolutely crucial, particularly to us, is that this new fine structure genetics is totally compatible with molecular genetics! Benzer is quite explicit on this matter: "Everything that we have learned about the genetic fine structure of T_4 phage is compatible with the Watson–Crick model of the DNA molecule" (Benzer, 1962, p. 83).

In short, what I would argue now that sweet harmony has been restored between the molecular and the biological is that the unqualified logical empiricist notion of reduction, that is one which sees a deductive relationship holding between theories, does indeed properly characterize the situation in genetics. The molecular revolution was one of reduction rather than replacement, although as we have seen, the full story is rather less straightforward and certainly rather more interesting than a totally ahistorical analysis would imply. It was not the older classical Mendelian genetics that was reduced (or which was open to reduction) but the newer fine structure or transmission genetics. What counts, however, is that the newcomer is just as biological as the oldtimer (Ruse, 1973a).

But what then is the present status of classical Mendelian genetics? My own feeling is that one can reasonably suggest that there is strong analogy between it and transmission genetics. Hence, were one to ask what its relationship to molecular genetics is, I would think it proper to argue that the relationship is one of weaker reduction, as characterized by Schaffner. To say that the relationship is one of replacement is to overstate matters and, as a matter of historical fact, to ignore significant continuities and links. Obviously I cannot expect the reader to accept such a claim as this about Mendelian genetics without any justification at all; however, I think the general plausibility of my position is fairly easy to show. It is agree by all that Mendelian genetics and transmission genetics are not identical — if they were, then there would be no argument! Mendelian genetics rolls together the units of function, mutation, and crossing over, whereas transmission genetics separates these out, allowing mutation of just part of the unit of function and crossing over within the unit of function. But where else are there essential differences? Just about everything else is retained from the old biological genetics in the new biological genetics, and what is crucially important, what differences there are do not make themselves much felt in most cases. Thus, the analogy between past and present is strong.

In support of what I am saying, take for example, the work of someone like Dobzhansky. The third and final edition of his classic *Genetics and the Origin of Species* appeared in 1951, before the work of Watson, Crick, or Benzer. In it he relied on the Mendelian law of segregation, its generalization to large groups (the Hardy–Weinberg law), and so on. The revised and retitled edition of his work *Genetics of the Evolutionary Process* appeared in 1970. The early chapters discuss, as accepted scientific fact, the major findings of molecular biology, for instance the Watson–Crick model of DNA. But what do we find then? All the old favourites like the Hardy–Weinberg law make their reappearance and play just as great a role in Dozhansky's theorizing as they ever did. Moreover, for page after page, chapter after chapter, the discussions in the old book and the new book are identical! Hence, not much change between the old and new seems to have occurred.

Of course, one might argue that Dobzhansky is holding to and working with two contradictory theories — traditional Mendelian genetics and modern molecular genetics. Thus, perhaps we should talk about the Hardy–Weinberg law$_M$ and the Hardy–Weinberg law$_m$. Alternatively one might say that, although the language is the same, the meanings of the biological parts of Dobzhansky's books differ, because the more recent book has replaced Mendelian genetics with the quite-different transmission genetics. Whereas previously we had the Hardy–Weinberg law$_M$ now we have the Hardy–Weinberg law$_t$. But I see no reason why one should assume either of these extremes. Dobzhansky explicitly identifies his unit of function (which he normally calls the 'gene') with the unit of function of transmission genetics, the cistron (Dobzhansky, 1970, p. 33). But, having made this change, he can immediately make use of just about all his old theory, because for the kinds of problems he tackles, the modern sophisticated analysis (replacing the old Mendelian gene) is not needed. Basically, Dobzhansky is untroubled by crossing over within the unit of function, because he works at a rather cruder level of analysis. Similarly, he just does not have to rethink all his work because of the Watson—Crick model. Hence, it seems to me by far the most reasonable thing to say that the modern *biological* geneticist incorporates into his new biological genetics much of the old Mendelian genetics. One simply does not find the tension one would expect were absolute replacement occurring. And this incidentally is why most evolutionists today can continue to talk cheerily of the importance of *Mendelian* genetics for their theory. The differences are really not so great that they feel the continuity is lost. After all, 'Darwinians' today hardly feel obliged to subscribe to many of Darwin's views, like his theory of pangenesis.

My overall position on the molecular revolution in genetics therefore is that, with respect to transmission genetics, we have strong reduction, and that with respect to Mendelian genetics we have weak reduction. Either way, the logical empiricist thesis does serve in highlighting the continuity between the biological level of understanding and the physico-chemical molecular level of understanding.

5.4. DAVID HULL OBJECTS

Just as we speak of a 'gaggle' of geese and 'murmuration' of starlings, so perhaps we should speak of an 'argument' of philosophers. At least, the reader who has followed my essays through to this point, will realize that no sooner has any philosopher made a claim, than a crowd of his/her fellows descend to demolish it. There is certainly something very Popperian about the way we all behave! But, painful though it may sometimes be, criticism is a good thing. Not only does it route out errors, but it helps both sides in a dispute to sharpen and articulate their own respective positions. This being so, I want now in this essay to acknowledge that one of today's most important philosophers of science, certainly today's leading philosopher of biology, David Hull, has with his customary wit and vigour, attacked generally and specifically the thesis I have just endorsed, namely that the logical empiricist notion of reduction appropriately and illuminately applies to the molecular revolution in genetics.[2] Scathingly, Hull states flatly: "I find the logical empiricist analysis of reduction inadequate at best, wrong-headed at worst."

And he adds that "the conclusion seems inescapable that the logical empiricist analysis of reduction is not very instructive in the cause of genetics. For my own part, I found that it hindered rather than facilitated understanding the relationship between Mendelian and molecular genetics" (Hull, 1974, p. 44. See also Hull, 1972; 1973b; 1976).

Hull has many disagreements with those people, like myself and Schaffner, who have endorsed the logical empiricist thesis. In this essay I shall concentrate on what I think is his most crucial, namely that there is something really rather irrelevant about the logical empiricist thesis of reduction, particularly as it is being employed toward an understanding of the revolution in genetics. Somehow the thesis directs us away from what is happening in genetics, rather than towards it (See Ruse, 1977b; Goosens, 1978, for other replies to Hull). Hull writes:

To my knowledge, no biologist is currently engaged in the attempt to reconstruct Mendelian and molecular genetics so that Mendelian genetics can be derived from molecular genetics. What is more, I can think of no reason to encourage a biologist to do so. This state of affairs strikes me as strange . . . Either this reduction is a peculiar case or else whatever it is that makes the pre-analytic notion of reduction seem important has dropped out of the logical empiricist analysis of reduction. (Hull, 1973b, p. 662.)

What makes Hull's objection particularly pertinent is that, even those who have endorsed the logical empiricist thesis, apparently agree fully with him about the actual place reduction seems to play in geneticists' research programmes. For instance, Schaffner concedes that the really exciting advances in molecular genetics have not come through the attempt to spell out a reduction, and he writes: "I wish to propose that the aim of reductionism is *peripheral* to molecular biology, and that an attempt to construe the development of molecular biology as exemplifying a research program or set of research programs whose conscious and constant *intent* is reductionism is both inappropriate and historically misleading" (Schaffner, 1974, p. 111). In a similar vein, one Michael Ruse thought it prudent to add the following footnote to one of his discussions of reduction in biology: "It is perhaps important to emphasize that no actual deduction from a purely molecular theory to a purely biological theory seems yet in existence. Indeed, reading the works of biologists, one gets the feeling that the whole question of reduction is more of a philosopher's problem than a scientist's" (Ruse, 1973a, 207n). What more damning comment could one make than that! Perhaps Hull is right and we should find better things to do with our time.

I have admitted already in this section that one of the great values of criticism is that, apart from revealing error, it can help one to articulate and understand more fully one's own position. We have here a case at issue. I confess that until Hull pointed the fact out, I really had no idea how muddled my thought on reduction in biology must have seemed. On the one hand, I was pushing a rather idealized picture of theory relationships; on the other hand, I was happily conceding that in real life scientists had neither desire nor intention to complete the overall scheme I endorsed![3] However, reflection has shown me that, although I need to eliminate the sense of paradox from my own position, this can be done without abandoning the logical empiricist cause and without denying the biological reality. In fact, thanks to Hull, I now feel even more convinced of the value of the logical empiricist thesis about theory reduction as it is applied to genetics. Let me offer two arguments why I think Hull's objection about the irrelevant, misleading nature of the logical empiricist thesis is not well taken.

First, although some of us philosophers may harbour a secret hope that what we produce will some day win us a Nobel prize in science, what we are supposed to do primarily is philosophy not science. Our job as philosophers is not to make brilliant new scientific discoveries, but to analyse and understand the way science works. In John Locke's great words, "it is ambition enough to be employed as an under-labourer in clearing the ground a little, and removing some of the rubbish that lies in the way to knowledge ... " (Locke 1959, Vol. 1, p. 14). Hence, that scientists are not pushing full-speed ahead on a reduction does not mean that philosophers ought not use the formal, idealized analyses of logical empiricism to see if they can throw light on the relationship between molecular and biological genetics. Scientists have their job to do, we have ours, and ours is certainly not to follow slavishly in the footsteps of science – nor is it necessarily a mark that we have done our job badly if scientists do not immediately rush to put flesh on our formal analyses.

Since this point I am trying to make is rather important, it may perhaps be worthwhile to consider for a moment a closely related example. No one could deny that in its present form, evolutionary theory taken as a whole is not a rigorously formulated hypothetico-deductive system. I have certainly not denied it in my earlier essays! Nevertheless, certain parts of evolutionary science, particularly those concerned with the spread of genes in populations, do approximate to a hypothetico-deductive theory and the philosopher can therefore use his model directly to throw light on these parts and on how these parts relate to other parts of evolutionary studies, like organic geographical distribution. (See Essays 1 and 2) Moreover, those places where the theory fails to fit the model can guide the philosopher to an understanding of the problems facing the evolutionist — essential data irretrievably lost, vast timespans, and so on. And although it is certainly true that logical empiricists would (or at least should) feel most uncomfortable were there no parts of evolutionary theory in any sense hypothetico-deductive or were there no sense in which evolutionary studies had progressed towards a greater manifestation of the hypothetico-deductive ideal — which incidentally they certainly have since the rather informal argumentation of the *Origin* (Ruse, 1975a), — there is nothing in their espousal of this ideal which insists that scientists must put this ideal before all else. It is fully recognized that things like missing data might make a significantly complete hypothetico-deductive evolutionary theory a practical impossibility, and that scientists might feel the technical details of filling out such a theory with full rigour to be rather boring and not something which would lead to dramatic new insights — as in mathematical proofs we often see how a problem can be solved but might not wish to fill in every last step.

An exactly analogous situation prevails with respect to theory reduction and genetics. Because of their understanding of DNA, the genetic code, and so on, biologists can see in broad outline how transmission genetics follows from molecular genetics in the manner suggested by logical empiricists, and as we have seen in the sickle-cell case, when it is in their interests they can often trace through the details of some particular case with precision. But for various reasons, particularly because it does not seem very exciting and would seem only to involve filling out details of a general picture already broadly understood, biologists are not trying to set up a fully articulated deduction of transmission genetics from molecular genetics. The logical empiricist, however, can and does recognize this fact. It is the broad outline which interests him/her — the general relationship between molecular and non-molecular genetics — and the way in which particular details are filled in when there is a specific need or interest. Hence, I would suggest that Hull is wrong when he writes that: "The crucial observation is that no geneticists to my knowledge are attempting to derive the principles of transmission genetics from those of molecular genetics. But according to the logical empiricist analysis of reduction, this is precisely what they should be doing" (Hull, 1974, p. 44). I do not think the logical empiricist analysis of reduction has this implication at all.

In a sense, however, one might feel that all of this is rather tangential. Surely I am missing the point. I am defending logical empiricism as a *philosophical* thesis. But is it not the case that Hull's main claim is not so much that the logical empiricist thesis on reduction is scientifically irrelevant, although he thinks it is, but that it is philosophically harmful? It 'hindered' his understanding of the true situation in genetics, rather than facilitated it (Hull, 1974, p. 44). In other words, it is no good defending logical empiricism as philosophy, because it is as philosophy that Hull is attacking it! It is here that I want to make my second argument for I think Hull is wrong in his claim. I would argue rather that the logical empiricist analysis does help us towards an understanding of what is really happening in genetics.

In order to defend the logical empiricist analysis, let me first ask of Hull the question: What, if the logical empiricist account is incorrect, is indeed the true situation? At times Hull seems to have no answer to this question, and seems almost in despair to give up hope of finding such an analysis. Rather, somewhat wearily, even he seems prepared to go along with the logical empiricist account.

If my estimation of the situation in genetics turns out to be accurate, then even the latest versions of the logical empiricist analysis of reduction are inadequate and must be improved. This conclusion, however, should not be taken as being too damning of the logical empiricist analysis of reduction. In spite of its shortcomings, it is currently the best analysis which we have of reduction. In fact, it is the only analysis which we have. (Hull, 1973b, p. 634.)

However, at other times he is more positive in his opposition. The logical empiricist sees the relationship between today's two theories of genetics as one of logical, deductive connection. The one theory, in some sense, 'contains' the other. Hull suggests, however, that the two theories belong in some important way to different worlds — at least, to incommensurable ways of viewing this world. He writes:

Most contemporary geneticists know both theories. They can operate successfully within the conceptual framework of each and even leap nimbly back and forth between the two disciplines, but they cannot specify how they accomplish this feat of conceptual gymnastics. Whatever connections there might be, they are subliminal. In a word, those geneticists who work both in Mendelian and molecular genetics are schizophrenic. The transitions which they make from one conceptual schema to the other are not so much inferences as *gestalt* shifts . . . (Hull, 1973b, p. 626.)

We seem, therefore, to have two sharply contrasting analyses of the situation in genetics. The logical empiricist like myself sees continuity. Talking of transmission genetics, we actually see this theory as being contained, or potentially contained, within molecular genetics; and even talking of classical Mendelian genetics we see strong links between the molecular and the biological. Hull, however, sees molecular and biological as separate. I confess that I am not quite sure whether by 'Mendelian genetics' in this passage, Hull is referring to classical Mendelian genetics or transmission genetics or both; but the point is not very important. Either way, Hull is denying a continuity endorsed by the logical empiricist. In fact, Hull implies not merely that molecular and biological are separate, but that they are separate in a very strong way. It is not so much that the molecular and the biological conflict, but that the different kinds of knowledge are about different kinds of being. The transition from one to the other requires a sharp break in focus — a 'gestalt' switch. To borrow an example used by the late Norwood Russell Hanson (among others), moving from one theory to the other requires the kind of mind-switch one makes when looking at drawings seen alternately as ducks or rabbits (Figure 5.8).

Admittedly, this is a somewhat extreme position I am ascribing to Hull, and it should, in fairness, be added that he does qualify the passage quoted above by saying: "To be sure, these observations are psychological in nature, and reduction as set out by philosophers of science is a logical relation between rational reconstructions of scientific theories . . . " On the other hand, he adds "but these psychological facts provide indirect evidence for the position I am about to urge concerning the logical relation". What is clear is that even if Hull would not accept all the implications that one might draw from the illustrations he uses — the kinds of radical implications that philosophers like Hanson (1958) who have relied on duck/rabbit illustrations have drawn about theory changes — Hull believes that, from a formal viewpoint, molecular genetics is a new theory replacing Mendelian genetics, not one absorbing Mendelian genetics: "If the logical empiricist analysis of reduction is correct, then Mendelian genetics cannot be reduced to molecular genetics. The long-awaited reduction of a biological theory to physics and chemistry turns out not to be a case of 'reduction' after all, but an example of replacement" (Hull, 1974a, p. 44).

Who is right? Reduction or replacement? Hull is certainly right in pointing to the way in which geneticists switch blithely to and fro between the molecular and biological levels, without giving too much thought to the relationship between them. But the extreme position to which he pushes this observation surely clouds rather than clears the view to correct understanding. Let us consider for a moment an actual situation in genetics where geneticists do work at both molecular and biological levels, to see which philosophical analysis throws the most light on the scientific practice and results. I take an example which has been much discussed by evolutionary geneticists recently,

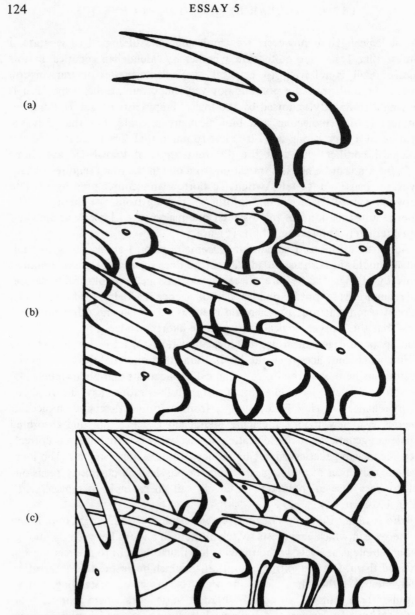

Fig. 5.8. Consider (a), What is it? Seen against a background of ducks, (b), one sees it as a duck. Seen against a background of antelopes (c) one sees it as an antelope. Hanson suggests that we get this kind of 'gestalt' switch, looking at the same thing in different ways, when considering facts against the background of theory (see Hanson 1958, pp. 13–14).

and which is in every respects a paradigm of molecular thought and biological thought coming up against each other: the problem of the amount of genetically-caused variation in natural populations. (The discussion which follows is heavily indebted to Lewontin (1974).)

There are, or rather were, two hypotheses about such variation when the problem was considered entirely at the biological level. On the one hand, followers of H. J. Muller argued for the so-called 'classical' hypothesis, namely that most organisms in a population are genetically similar, with just the occasional mutant allele, usually recessive, usually deleterious, deviating from the 'wild-type' (Muller, 1950). On the other hand, followers of Th. Dobzhansky supported the 'balance' hypothesis (Dobzhansky, 1951). They saw selection primarily maintaining genetic variability in a population, for example, through superior heterozygote fitness, as in the sickle-cell case: at equilibrium both alleles are retained in a balance. Hence, for balance hypothesis supporters, there is in an important sense, no such thing as a wild-type: almost every genotype in a population is different. Diagramatically we can represent the differences between the two populations as shown in Figure 5.9.

$$\frac{+ + + + m + \ldots + + +}{+ + + + + + \ldots + + +} \qquad \frac{+ + + + + + \ldots + m +}{+ + + + + + \ldots + + +}$$

(a)

$$\frac{A_3\, B_2\, C_2\, D\, E_5 \ldots Z_2}{A_1\, B_7\, C_2\, D\, E_2 \ldots Z_3} \qquad \frac{A_2\, B_4\, C_1\, D\, E_2 \ldots Z_1}{A_3\, B_5\, C_2\, D\, E_3 \ldots Z_1}$$

(b)

Fig. 5.9. The difference between the classical and balance hypotheses. (a) shows the classical hypothesis. Apart from the occasional deleterious mutant 'm' any two individuals will have identical corresponding so-called 'wild n type' genes, '+'. (Obviously the genes at different loci are not the same.) (b) shows the balance hypothesis. One expects a range of alleles at any locus, individuals are often heterozygous, and with respect to any locus, two individuals are usually different (from Lewontin, 1974, pp. 24 and 25).

Now, at the biological level, it is all but impossible to decide between these hypotheses. Balance hypothesis supporters argued that the phenotypic differences caused by allelic changes would normally be very slight, although such differences could have significant evolutionary implications. But the measurement of slight differences due to genetic change is incredibly difficult, mainly because environmental changes can also cause phenotypic differences and one can never be absolutely sure that one is keeping the environment constant. Remember, for instance, the environment of an individual includes the other members of its population.

What geneticists have done in the past ten years, however, is to switch the debate from the biological to the molecular level. As we know, functional strips of DNA lead, via RNA, to polypeptide chains — strings of linked amino acids. Changes in DNA lead (taking note of redundancies) to changes in the polypeptide chains. Using a technique known as 'gel electrophoresis', which is based on the fact that some amino acids have different electrostatic charges and that, hence, changes in polypeptide chains lead to changes in charges, geneticists can detect changes in polypeptide chains, which, in turn, they interpret as changes in the DNA structure (see Ayala and Valentine (1979) for details).

Both classical and balance hypothesis supporters think that the findings about polypeptide chains are relevant to their debate. In particular, it is found that polypeptide chains (caused by DNA strips at the same locus) do vary greatly. In other words, there seems to be a great deal of genetic variation in populations of organisms (Table 5.1). In humans, for instance, we have about 100 000 structural gene loci and an average heterozygosity of 6.7%. This means that any human individual is heterozygous at about 6700 structural gene loci! (Ayala and Valentine, 1979, p. 81). Somewhat naturally therefore, balance hypothesis supporters take these and like facts to be confirmation of their position, namely that selection maintains genetic variability in populations. However, we find that scientists, no less than philosophers, are good at defending their positions when under attack. Classical supporters, whilst they cannot deny all of this variation, argue that the variation is nonadaptive, that it is due to drift not selection. And one must, in fairness, note that classical hypothesis supporters can give both theoretical and empirical arguments to support their position (see, for instance, Kimura and Ohta (1971) and Lewontin (1974) for details).

Clearly, the resort to the molecular level has not brought debate to an end — if anything, it rages more fiercely than ever before. But what I want to ask is what is the most reasonable way to interpret geneticists' resort to the molecular level at all? Why did geneticists get so excited about gel electrophoresis and so on? On Hull's interpretation of the situation it is difficult to see why geneticists should have turned, in the instance we are considering, to the molecular level in the first place. Why did they think, as they obviously

TABLE 5.1
Genetic variation in different groups of animals and plants
(from Ayala and Valentine 1979, p. 81)

Organisms	Number of species studied	Average number of loci studied per species	Proportion of polymorphic loci per population*	Proportion of heterozygous loci per individual
Invertebrates:				
Drosophila	28	24	0.529	0.150
Wasps	6	15	0.243	0.062
Other insects	4	18	0.531	0.151
Marine	14	23	0.439	0.124
Land snails	5	18	0.437	0.150
Vertebrates:				
Fish	14	21	0.306	0.078
Amphibians	11	22	0.336	0.082
Reptiles	9	21	0.231	0.047
Birds	4	19	0.145	0.042
Mammals	30	28	0.206	0.051
Average values:				
Invertebrates	57	21.8	0.469	0.134
Vertebrates	68	24.1	0.247	0.060
Plants	8	8	0.464	0.170

*The criterion of polymorphism is not the same for all species.

did, that the molecular level was going to break down some barriers? Given Hull's analysis, geneticists had a debate at the biological level. It was not getting untangled. So they gave up, made a gestalt switch, and turned to a different problem, a problem at the molecular level. Two levels, two problems, and never the twain shall meet. On the logical empiricist analysis, however, all is readily explicable, including geneticists' enthusiasm for the molecular level. Mendelian genes (or cistrons) can be identified with strips of DNA, DNA causes polypeptide chains, and these either lead to, or can be identified with, gross phenotypic characteristics. Hence, inasmuch as one gathers information about genetically caused variation amongst polypeptide chains, one is throwing light on genetically caused variations at the gross phenotypic level – the matter at the heart of the debate in the first place.

Thus, what we find in the case of balance hypothesis supporters, is the belief that much of the difference in the polypeptide chains reflects as difference at the gross phenotypic level – difference which they believe must be maintained by selection. And in support of their case they point to several facts which they believe explicable only on their hypothesis; facts, at least, which are certainly not explicable on the classical hypothesis. One thing in particular they believe proves their case, namely that polypeptide ratios between closely related groups are often very similar, pointing to a systematic cause, which they believe can only be selection and is certainly not drift. Obviously this selection works on the phenotype and, hence, equally obviously the molecular is being related to the biological (Ayala *et al.*, 1974).

But the position of classical hypothesis supporters is also readily explicable given the logical empiricist analysis. They cannot deny the differences in polypeptide chains; however, they argue that either different chains lead to the same gross phenotypic effects, or that different chains lead to different effects but that these effects are selectively neutral between themselves (see, for instance, King and Jukes, 1969). But either way, we see that they do believe that polypeptides have direct implications for gross phenotypic differences, which is what the logical empiricist account leads us to expect, and indeed demands. If they do not believe this, why do they not simply ignore the results of gel electrophorectic studies as irrelevant?

What more can one say — or want to say? As I see things, if one takes Hull's position seriously, then the actions of evolutionary geneticists over the past fifteen years or so really do not make much sense. Not only are they working on quite separate sets of problems, but mistakenly they think that there are connections between the sets. On the logical empiricist analysis, however, what geneticists have been doing and are still doing makes perfectly good sense. Their actions and enthusiasm are quite understandable. And this is about as good a recommendation as one could have for a philosophical thesis. It may not tell us what scientists are going to do, or even what they should do, although I can imagine times when a philosophical thesis might be prescriptive. But if a philosophical thesis, such as the logical empiricist thesis applied to genetics, tells why scientists do as they do do, then I for one feel that philosophy has fulfilled its proper function. For this reason I argue, and I think it is worth arguing, that the molecular revolution in genetics was one of reduction and not of replacement.

NOTES

[1] For the sake of clarity I am blurring over some matters here, for instance that a piece of DNA might have a function which would not be reflected immediately at the pheotypic level and, therefore, might not be thought of as having an equivalent Mendelian

gene. Although things like these will make matters a great deal more complicated, I doubt they will affect he central truth of my claims.

[2] From now on, I shall speak simply of 'reduction', no longer feeling always obliged to make the exact distinctions I drew at the end of the last section.

[3] Perhaps my concession was not that happy. My experience is that when philosophers put things in footnotes, they are often glossing over matters about which they feel slightly guilty.

DOES GENETIC COUNSELLING REALLY RAISE THE QUALITY OF LIFE?

When I was researching to write a book about Charles Darwin and the great nineteenth-century debate about evolutionary theory, one rather tangential, pathetic fact struck somewhat more deeply than I might have expected. Many of the major figures in the Darwinian Revolution had lost children through one or other of the dreadful diseases which so scourged Victorian nurseries. Darwin himself lost two children, including his favourite, Annie — a loss which he and his wife took sadly to their graves (Darwin, 1887). Huxley too, felt such a burden of grief, when his first-born, Noel, died suddenly at the age of four: "He was attacked with a bad form of scarlet fever on Thursday night, and on Saturday night effusion on the brain set in suddenly and carried him off in a couple of hours" (Huxley, 1900, Vol. 1, p. 229). And Darwin's other great supporter, Joseph Dalton Hooker, also lost a child, as did others. When, as we are wont to do, we poke fun at the Victorians' obsession with death and funerals, particularly those of small children, we should remember how ever present these phenomena were to them (Morley, 1971).

But why do we have to make such an effort? Obviously because things have changed so much. Scarlet fever, tuberculosis, diphtheria, small-pox, typhoid: where are they today? How many of your colleagues have lost a child to one of these? None of mine have, and I work in two departments with a combined total of nearly fifty people. Thanks to modern medicine these terrible illnesses threaten us no longer, nor even do somewhat milder diseases like measles and whooping cough. In other areas also, medicine has made fabulous advances in this century, since the war even, and the quality of life has been elevated significantly for nearly all people — if not for each one of us directly, then for someone near us. People with terrible burns or with broken bodies are brought back from the brink of death. Failing hearts are revived and given new lease through pace-makers and bypasses, and the like. Cripples are straightened and the lame literally made to walk again. Admittedly not all is triumph, at least not yet. People still die of cancer, despite significant progress with radiation treatment and chemotherapy. And although we seem to have the technological ability to perform heart-transplants, difficulties due to rejection remain to be solved. But these are problems at the frontier, and certainly do not belittle the very great advances that have been made.

In recent years one much-publicized advance in modern medicine has involved the increasingly widespread practice of so-called 'genetic counselling'. In various ways, through identification, contraception, sterilization, and abortion, the attempt is being made to minimize the birth of people with severely deleterious genes, sufferers from 'genetic disease' (Birch and Abrecht, 1975; Hilton *et al.*, 1973; Epstein and Golbus, 1977; Scrimgeour, 1978). One centre for genetic counselling in London England, for instance, which has been in existence for rather less than twenty years, is now giving advice and help to around 2000 families per year (Polani *et al.*, 1979). And the practice is increasing, in N. America, Europe, and elsewhere, as medical techniques get more sophisticated and as the supposed virtues and benefits of genetic counselling get ever-wider publicity (Webb *et al.*, 1980; Simpson *et al.*, 1976; Galjaard, 1976; Gershon *et al.*, 1977).

Prima facie, one might think there could be no argument that genetic counselling is a good thing, or, to use a popular modern phrase, something which raises the quality of life. I am sure that many of us know of families where one or more of the children suffer from Down's Syndrome ('mongolism'), and we know what a drain, physically and financially, such children are on the parents — and this is not to mention stresses on the siblings. Can there be any doubt that one would want to avoid such a human tragedy? I ask this question, notwithstanding the fact that people with Down's Syndrome are very loving and cheerful people. And who could fail to want to prevent the birth of a child with sickle-cell anaemia, condemned to an early death? Or what about the unfortunate who bleeds at the slightest cut, doomed to years of crippling pain at the body joints? If we can prevent these and like afflictions, then surely it is right to do so, just as it was right for earlier generations to attempt to prevent tuberculosis and smallpox.

But is this really so? Let me say simply that I do not find matters quite so straight-forward as first appearances might incline one to think, and I should like to share with the reader some of the reasons for my hesitancy in giving genetic counselling unqualified endorsement. I assume that ultimately the key factors in evaluating the worth of something like genetic counselling are moral, whether one asks if it is a good thing, or desirable, or if one queries if it raises the 'quality of life' or whatever other trendy term one might like to use (Michalos, 1976; McCall, 1976). This being so, as in all moral issues, matters seem to come down to two prime factors and their balance.

First, one has the question of happiness, particularly general happiness.

This is the so-called 'utilitarian' element to moral issues (Smart 1967). If we are to call something like genetic counselling a good thing, then we expect to see happiness levels raised, and conversely we expect to see unhappiness levels decreased. Drains are a good thing, they raise the quality of life, because we are all happier if wastes are carried away efficiently. Of course, general happiness might be raised, even though one only shares in things potentially. Most utilitarians would prefer to argue that the happiness of others should be sufficient to make one judge something to be a good, but even if one insists on relating things selfishly to one's own interests, a case can be made easily for the value of facilities ready for all. I may never need the emergency room in a hospital, but as a utilitarian, one can judge such a facility to be a good thing, both because one dislikes seeing others suffer and because it is there if ever one needs it oneself. The peace of mind that is brought by the knowledge that there are adequate medical facilities available is undoubtedly a good.

But there is a second side to morality. Counterbalancing utilitarianism with its emphasis on happiness, particularly general happiness, one has the question of the rights of the individual: what might be called after the greatest of all the German philosophers, the 'Kantian' element to morality (Körner, 1955; Beck, 1960). Speaking about a society or a group, and obviously in considering genetic counselling, we are looking at it from a group perspective, in order to make things better or to have something which can be labelled a 'good' or in order to achieve high quality of life, one must treat people as ends not means; one cannot blithely sacrifice individuals for the sake of the common good or happiness. Unfortunately, this is not a perfect world, and so sometimes one must take actions clearly for the overall happiness even though certain individuals benefit but little; but no society has a high quality of life despite a general happiness if it is purchased only at the abject misery of a minority — particularly if this minority is made up of innocents and not those with idiosyncratic group-happiness destroying propensities, like rapists. Something is not a good thing if some suffer for the sake of others, even if the others are a majority.

Against this ethical background, let us turn now to some of the facts about genetic counselling and to the positive case that can be made for it. Then some queries can be raised.

6.1. GENETIC COUNSELLING

Most conceptions resulting in a fetus that might be termed genetically 'defective' abort spontaneously; those that do not, result in some 2–4% of live births (Scrimgeour, 1978). They fall into three kinds: chromosomal abnormalities, ranging from the very severe, like various form of trisomy leading to Down's syndrome (Figure 6.1), to the less severe, like duplication of the Y

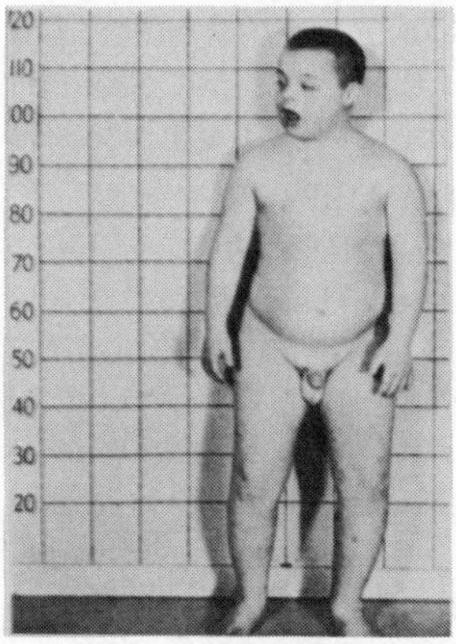

(a) 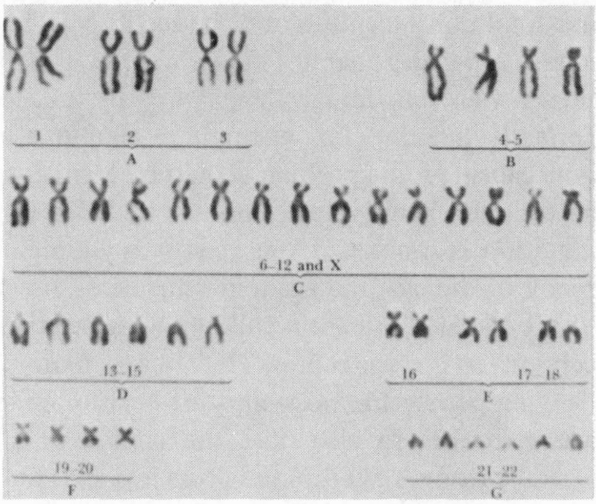   (b)

Fig. 6.1. (a) A child with the characteristic features of Down's Syndrome and (b) the typical 'karyotype', set of chromosomes. Humans normally have 46 chromosomes; people with Down's Syndrome have an extra chromosome, which in this case occurs in the G group (from Rothwell, 1979, p. 276).

chromosome in males which is linked to an increased tendency in anti-social behaviour (Borgaonkar and Shah, 1974); abnormalities of single genes, causing ailments like cystic fibrosis and haemophilia; and multifactorial disorders, where not only several genes but also the environment seem involved, causing such problems as spina bifida, congenital heart defect, and club feet (Epstein and Golbus, 1977; Galjaard, 1978; Passarge, 1978).

Now, the problem at issue is how one can avoid the births of people thus affected, and as a first step one must obviously ask how it is that one might get such a genetic defect. (I realize that in talking of a genetic 'defect' I am ignoring a host of problems. These are being shelved but not forgotten.) Causally speaking, there seem to be two potential reasons for such defects. Either one's parents (one or both) might themselves have the defect, or at least have genes which can cause the defect (i.e., they could be carrying the genes recessively) and these are then passed on, or the defect might be caused by a mutation in the sex-cells contributed by one's parents. Taking these disjuncts in turn, certain remedies seem obvious. Clearly if potential parents themselves have a genetic defect or, if in the parents, one can identify genes with the potential for defect, then one way to avoid genetically-handicapped offspring is for the would-be parents not to conceive at all — that is, to use total contraception or to be sterilized. This course of action obviously demands a penalty, namely that a couple will then have no natural children of their own, even though the chances are that some of their offspring would be perfectly healthy. For example, a couple, each heterozygous for a recessive gene causing some problem, could by Mendel's first law expect, on average, to have three phenotypically normal children for every crippled child. As far as mutation is concerned, one cannot as yet identify problems at the level of the sperm or the ova, but again in some cases one can identify an 'at risk' segment of the society, and again children with defects can be avoided through contraception and sterilization. Here, even more than last time though, one is denying people the possibility of healthy children. For instance, as maternal age increases, so also does the chance of a child with Down's syndrome. Overall, Down's syndrome occurs in 1 in 600—700 live births. But in mothers of 40 the occurrence is 1 in 50—100. One can clearly reduce the incidence of Down's syndrome by advising women of 40 not to reproduce, but at the same time one is denying 98 out of 100 mothers of 40 the opportunity of having healthy children (Epstein and Golbus, 1977; Passarge, 1978. See also Figure 6.2).

The breakthrough in genetic counselling has been the development of techniques for studying the nature of the fetus, that is, the child before birth.

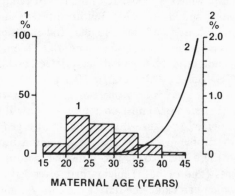

Fig. 6.2. Chart comparing maternal age groups and proportion of total birth rate (1) against the incidence of Down's Syndrome (2) (redrawn from Passarge, 1978, p. 27).

If the fetus is deformed then it can be aborted and the prospective parents then have the option of avoiding further conceptions or trying again for a healthy fetus. One very powerful technique for the examination of the fetus involves ultrasonic scanning, a process similar to radar, which uses reflected sound waves to measure distances and densities, and which can provide incredibly informative pictures of the situation within the womb (Donald, 1978) (Fig. 6.3). Obviously however, such scanning has its limits. It could not, for instance, tell one about a genetic ailment which manifests itself sometime after birth. Here, however, another technique is available: *amniocentesis*. This is a process whereby fluid surrounding a fetus — the 'amniotic' fluid — is sampled, via a needle inserted through the abdomen, and then the fetal cells in this fluid are studied. One can do direct chromosomal tests, or culture the cells and study for biochemical abnormalities. Thus, for instance, one can tell if a fetus has Down's syndrome, because it will then have an extra chromosome; similarly one can tell if the fetus has certain forms of genetically caused biochemical defect (Bartsch, 1978; Epstein and Golbus, 1977).

Prima facie, the benefits to potential parents brought about by genetic counselling using ultransonic scanning, amniocentesis, and so forth are immense. They no longer face being burdened with a child whose very existence will cause them strain and unhappiness. Moreover, in order to avoid such a calamity they will not be required to sacrifice the opportunity of their own, natural, healthy children. At the level of the individual therefore, new developments in genetic counselling seem very great. But, since morality transcends the individual, going up to the group, let us cast our nets somewhat wider in looking for benefits. Hopefully, all in our society can have the opportunity to draw on genetic counselling if need be, not just the very rich or very priviledged — so it has a general application in this respect. Genetic counselling seems to be a good thing from the group perspective. It raises the general quality of life. And this seems to be so indirectly as well as directly. Fairly straightforward cost-benefit analyses apparently show that money and resources are saved through such counselling — money and resources which might well be used elsewhere to improve health care and so forth (Glass and Cove, 1978; Mikkelsen *et al.*, 1978; Hagard *et al.*, 1976) (Fig. 6.4).

In order to put some flesh on this rather abstract discussion, let us look briefly at some relatively early reports of one of the most complete programmes aimed at a particular genetic disease. It illustrates well genetic counselling in action, as well as giving some insight into costs versus benefits. (Kaback and Zeiger (1973), Kaback *et al.* (1974), Childs *et al.* (1976 a, b). See also Brock (1977, 1978) for details of a programme for a different disease.)

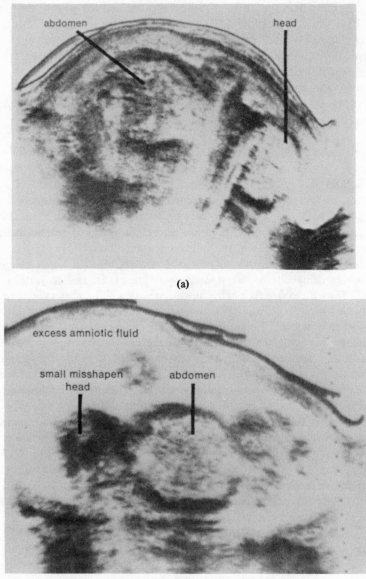

(a)

(b)

Fig. 6.3. Ultrasonic scanning, which uses sound to visualize fetuses, produced these two 'sonograms'. That above is of a normal pregnancy, and that below is of an anencephalic fetus with misformed head (from Epstein and Globus, 1977, p. 705).

Fig. 6.4. Genetic disease can affect the famous and talented just as much as anyone else. It used to be thought that Toulouse-Lautrec was an *achondroplastic dwarf*, caused by a new mutation. However recently it has been suggested that his characteristics better fit the syndrome known as *pyknodysostosis*. Amongst the features associated with this syndrome, and possessed by Toulouse-Lautrec, are fragile bones (he broke his thighs twice), shortened stature (he was about 5 feet tall), large skull and receding chin (he grew a beard to conceal the latter), and shortened fingers and toes (one acquaintance spoke of his 'comical little hands'). Pyknodysostosis is caused by homozygous possession of a particular recessive gene. Toulouse-Lautrec's parents were first cousins, which inbreeding could well have made such a gene homozygous.

6.2. THE JOHN F. KENNEDY INSTITUTE TAY-SACHS PROGRAMME

Tay-Sachs disease is a neuro-degenerative disorder, caused by abnormal storage of the sphingolipid, GMZ ganglioside, in the central nervous system and elsewhere in the body. It is invariably fatal before the age of five, and what makes it perculiarly vile is that the affected children appear quite normal until about six months, after which they rapidly collapse into a zombie-like trance, and soon die — often of a respiratory failure. It is a recessive condition, that is, the diseased children are homozygotes for a particular recessive gene, and each parent is a heterozygote for this gene. However, although at the noticable level heterozygotes are phenotypically normal, biochemical tests can identify them, and through amniocentesis and cell culture the same can be done for affected fetuses. Fortunately, on average Tay-Sachs disease is not very common. Without action, one would expect about 50 affected new-borns per year in North America. What makes Tay-Sachs disease particularly suitable for a mass screening programme is that nearly all of these affected new-borns (45—50%) occur in one readily identifiable segment of society: Ashkenazi Jews.[1] An Ashkenazi Jew has a 1 in 30 chance of being a heterozygote: hence there is a 1 in 900 chance that a couple will both be heterozygotes, and because there is a 1 in 4 chance that any one of their children will be affected, this means that there is a 1 in 3600 chance that an Ashkenazi Jew new-born will be affected. In other words, in Tay-Sachs one not only has a disease which is discoverable both potentially and actually, but one knows that essentially it is confined to one identifable, relatively small portion of the population at large.

Armed with these facts, members of the John F. Kennedy Institute (out of Johns Hopkins University in Baltimore) set up a mass screening programme for high-risk populations in Baltimore and Washington. Acting through the medical community and even more particularly through the religious community — first the rabbis and then various lay groups, especially those for women — they tested people (mainly couples) to see if they were carriers and, where appropriate, to see if fetuses in early pregnancy were affected. In the first year of the programme, understanding an 'at risk' couple to be a couple where both partners are heterozygotes, the following figures were obtained:

Population	Results
Subjects tested	6938
Carriers identified	315
'At-risk' couples identified	11

None of the couples 'at-risk' had, as yet, had affected children (3 of the couples had 2 children and one had 1 child), and only 4 had any family history of Tay-Sachs disease. Following identification, five pregnancies occurred in the couples, and one fetus in a couple with no family history of Tay-Sachs disease was found to be affected and aborted.

All of this effort to find one affected fetus may seem like a case of excess, although note that this was an early report on an on-going programme. (This was a report published in 1974.) But on the basis of their experience, the investigators felt able to draw up a fairly comprehensive cost-benefit analysis. First, there was the matter of the costs:

1. Costs are calculated on basis of testing 50 000 people over 3 years.

2. Costs include costs of leukocyte retesting in approximately 6% of all subjects screened (mainly serum inconclusives and females on BCP).

3. Costs calculated by assuming each person's serum will be assayed an average of two times.

4. Exclude professional salary, overhead, depreciation, etc.

   Program cost for 3 years:

Equipment	30 000
Technicians (3)	81 000
Coordinator and secretary	50 000
Lab and office supplies	15 000
Printing, mailing, etc.	10 000
	$186 000

Retesting (WbC) costs − 3000 @ $3 per person	9 000
	$195 000

   $195 000 − 3 years = $65 000 /yr.

240 work days/year; 200 assays/day × 150 days = 30 000 tests/year or 15 000 people screened annually.

$$\frac{\$65\ 000}{15\ 000\ \text{persons}} = \$4.35 \text{ per person screened}$$

(Kaback et al., 1974, p. 129)

Then, assuming that there are some six million Jews in North America, one can compare the total cost of screening those most likely to have children (married people between 18 and 43) and the cost of the affected children they might have without screening (and abortion of the affected).

Service	Subjects	Costs
Screening tests	1.5 × 10 18 to 43-yr-old married subjects @ $4.00	$6.0 × 10^6
Amniocentesis, cell testing	850 couples; 2267 pregnancies @ $100.00	$0.23 × 10^6
Abortions	567 pregnancies @ $200.00	$0.12 × 10^6
Total cost of program		$6.35 × 10^6
Average medical cost/yr/child (over 3.5 yr)	425 children with Tay-Sachs disease (life expectancy: 3.5 yr)	$15 000.00
Average lifetime medical costs/ child		$52 500.00
Total cost of 425 children		$22.3 × 10^6

(Kaback *et al*., 1974, p. 130)

As can be seen, the figures alone make a strong *prima facie* case for such a screening programme (6.35×10^6 as opposed to 22.3×10^6). And this is not even to think of the human tragedies of couples who have Tay-Sachs children. In short, we seem here to have a paradigm for modern medicine at its best. In the terms of this essay, who could deny that this programme, and those like it, significantly raise the overall quality of life?

6.3. THE LIMITATIONS TO GENETIC COUNSELLING

The time has now come to consider the other side of the coin. All too frequently, what appear at first to be radical forward steps due to new technology — medical or otherwise — soon are found to be both limited and bearers of unwanted side-effects. The use of DDT to combat mosquitoes and thus malaria is a paradigm, but it is by no means the only example which springs readily to mind. First, in looking at what we might term the 'darker side' to genetic counselling, let us ask about its limitations. What is genetic counselling *not* going to do about genetic disease?

Well, one thing that genetic counselling is not going to do is eliminate all genetic disease. In one sense, of course, this is not that worrisome, for it is by no means obvious that, even if one could so eliminate all genetic disease, one would really want to. Some genetic ailments are readily curable by conventional means and one might, with reason, decide that one would rather do this than either abstain from having children altogether or go the route of amniocentesis and possible abortion. Pyloric stenosis, for instance, causing a blockage from the stomach to intestine, is genetic but curable by a relatively straightforward operation (Crow, 1973).

But obviously there are some genetic ailments one would like to eliminate, in the sense of not having any such affected persons born — conventional cure does not exist and seems highly unobtainable. However, for a number of reasons, total elimination through counselling is far from possible — indeed would at least cause more problems than it cures. For a start, there is the problem of whom one is going to counsel. Suppose one decided to set about matters in the most radical way possible and that one was going to counsel every pregnant woman: one hoped to see that no fetus survived unless it were genetically perfect. Now, leaving for the moment the whole question of those women who did not very much want to be so counselled, and leaving also the women who somehow avoid such counselling, it is clear that a total counselling programme aimed at all genetic disease, would necessarily demand that amniocentesis be performed on every pregnant woman. Otherwise, one could not hope to find new instances of genetic disease caused by new mutations, and undoubtedly just by checking family histories one would miss many cases of genetic diseases, the potentials for which are being passed on recessively. Unfortunately, this kind of universal amniocentesis in itself raises problems.

First, one can only tap a pregnant woman for so much amniotic fluid (Bartsch, 1978; Bowser-Riley, 1978). But, to check for all possible diseases one would literally need gallons! Ever-refined microscopic methods are being devised and employed, but there are limits to making all tests possible. Hence, we have a limitation here. Second we have a limitation in that no one can (or could) say absolutely that amniocentesis is totally safe. Again, practice and refinements are making it ever safer, but not even its most enthusiastic proponents are prepared to allow that the risk to the fetus (in unwanted abortion or damage) will drop much below 0.1 to 0.2% (Epstein and Golbus, 1977; Turnbull, 1978; NICHD, 1976; Simpson *et al.*, 1976). And this is probably a highly optimistic estimate, based on work done (as it is today) in leading medical centers and hospitals. Other figures that one sees quoted put the risk from amniocentesis at at least an order of magnitude higher, that is at 1–2% (MRC, 1978). If mass screening is to take place, something of the order of a Pap smear, then probably we would have the prospect of the local GP doing amniocentesis in his/her office, and one doubts that under those circumstances the risk would even stay as low as it presently is. But with risk figures as high as they are, given the fact that technologically we can test for a limited number of ailments, the whole desirability of mass screening programmes is thrown into question. Suppose it be suggested, as has indeed been suggested, that a mass screening programme for Down's Syndrome be set up (Stein *et al.*, 1973). That is to say, instead of trying to eliminate all genetic diseases, we try to eliminate just one. The problem is that the risk of such an affected child is, on average, about 0.2% — the very minimum figure mooted as the risk from amniocentesis! In other words, a mass screening programme, even restricted to one ailment like Down's Syndrome, in a sense may well cause as much damage as any possible good it can do. And if one takes some of the higher figures for amniocentesis risk, one may, in fact, be destroying 10 healthy fetuses in order to find any one Down's syndrome fetus. One may think the price worth paying; but it is a price.[2]

The third objection to a mass screening programme aiming at all genetic disease is that, as yet, we cannot test for all genetic diseases (Epstein and Golbus 1977; Brock, 1977). Moreover, the hope of ever so doing for many of the diseases seems remote indeed. "It is likely that many of the autosomal dominantly inherited diseases will not be identifiable by an abnormal gene product" (Epstein and Golbus, 1977, p. 710). There are, indeed, ways in which one can try to tackle problems such as these — for instance, if one can find another gene closely associated ('linked') with the deleterious gene, one can try to discover this 'marker' gene, and thus infer the presence of the deleterious gene. However, approaches like these still, as yet, lie much in the realm of theory and, in any case, cannot hope to be absolutely foolproof. The linkage between genes is always snapped sometimes (see Essay 5).

For these various reasons therefore, hopes of total elimination of all genetic diseases, or even of the total elimination of some diseases, seems unattainable — and even if that goal were attainable, the 'cure' might be worse than the disease, given the cost in lost and damaged fetuses. Nor could one decide to accept the unpleasant side effects in a one-shot effort to stamp out genetic disease: that one would balance the cost in lost and damaged fetuses for one generation in order to eliminate genetic disease for all future generations. Apart from anything else, new mutations make this a vain hope. Indeed, one suspects that considering all of the additives that we put in our foods and the undoubted mutagenic effect of some, if we were really keen to cut down on genetic disease we could more profitably turn to unpolluted food rather than all-encompassing screening programmes (Evans 1978). Hair dye has also been implicated in genetic defects — hair dye!

Of course, nothing that has been said in this section denies the value of genetic counselling for individuals or groups who can be identified as having high risks of serious genetic disease, nor has it been intended to. For instance, it has been mentioned that older mothers stand a much higher chance than normal of having mongoloid children. In a case like this, one might well decide that the benefits of counselling including amniocentesis significantly outweigh the costs. Nor is anything which has been said been intended to downplay the very great success of those involved in medical genetics at extending the scope of their theories and techniques to discover an even wider range of ailments. In recent years much successful effort has been expended on the detection of so-called 'neural-tube abnormalities'. These involve defects in the neural tube and can cause horrendous difficulties in functioning. One such defect, spina bifida, means a life-time's confinement to a wheelchair, and total incontinence, just to mention two of its vile aspects (Brock, 1977; 1978). The point that I have tried to establish in this section is the *even if we were fully prepared to abort every diseased fetus*, genetic disease as a continuing threat and reality is not going to go away. Indeed, if anything as a threat, genetic disease is going to increase slightly because now at-risk couples will not abstain entirely from natural children, but will persist, hoping to get healthy children even though some of these will, in turn, be carriers of the disease (Kaback *et al.*, 1974).

6.4. THE PROBLEM OF ABORTION

But there's the rub! "Even if we were fully prepared to abort every diseased fetus." So far, I have been assuming that the whole question of abortion has no bearing on the moral status of genetic counselling. But obviously this is an assumption that cannot go unquestioned. There are two parties immediately concerned in the problem of abortions: the fetuses and the parents. Let us take them in turn.

A Catholic or like thinker would, of course, deny absolutely that abortion can ever be justified. To him (or her) a claim that an abortion could be in a fetus's moral interests would be false or, more probably, meaningless. To me, matters are not quite so clear-cut, although I do admit there is something a little odd about the whole discussion. I certainly share with the Catholic the belief that the fetus is more than just a blob — a 'tumor', with pregnancy a form of venereal disease, one forthright participant in the genetic counselling debate has called it (Hilton *et al.*, 1973, p. 212) — although I am not sure that I would go all the way with the Catholic and afford the fetus full personhood.[3]

But, lacking insight into the ultimate cosmic purpose to life, and recognizing also that morality would seem to involve minimizing unhappiness as well as maximizing happiness, I suspect that even viewed from the fetus's view-point, some case can be made for the abortion of the severely diseased in the hope of raising quality of life or otherwise making things better. I cannot see that a life of genetically caused extreme pain is a life worth living, and I would not much want to live my life if it caused suffering to others, even though I myself did not much suffer. Of course, this decision to live or not to live is one which must be made for the fetus, but this is no different from decisions which must be made for children and incapacitated adults. Hence, from the fetus's viewpoint, inasmuch as effects of this counselling might be open to all of us, the quality of life seems improved. In other words, in this respect a case can be made for saying that genetic counselling is a good thing or raises the quality of life.

But there are at least two problems which must be mentioned. First, it must be recognized that genetic counselling could well increase the psychological agony of those who are born with genetic disease. Suppose one has haemophilia. That in itself would be bad enough. But then add to it the knowledge that one's parents had had aborted several of one's younger siblings, with the advice and help of the best of the medical fraternity. I am not saying that this in itself is a reason not to counsel. I do rather suspect, however, that the knowledge that one's parents would have had one aborted if they could is not exactly something which would improve one's self-esteem, something already probably battered by one's ailment.

Second, even more seriously, there is the fact that some genetic counselling today involves the destruction of the fit as well as the unfit. This applies particularly to sex-linked characteristics, like haemophilia. One cannot tell whether a particular child is a haemophiliac, but one can avoid the birth of haemophiliacs by aborting all male fetuses simply because the disease only exhibits itself in males. The penalty one has to pay, however, is the abortion of healthy males (some 50% of all the males). Morever, of the non-aborted females some 50% will be carriers of the haemophiliac gene: 50% of their sons would be haemophiliacs.[4] Obviously, inasmuch as one is aborting healthy males, one is hardly involved in a practice which can in any way be described as a particularly good thing. If one takes seriously the notion that goodness cannot be brought about in the sense of increasing happiness if it is purchased at the expense of the happiness of others (or that only very rarely can this happen), then it seems questionable that this kind of counselling raises the overall quality of life or any such thing. No doubt how one will feel on this matter rests partially on the degree to which one argues that the aborted male fetus is a person, with rights. But, even if one does not accept it as a full human, one might argue that the desire of people to have their own natural children does not override the right of a healthy fetus to be born. I recognize of course that many people will not share my concern here. So be it. I think that other arguments I am giving in this essay should still make them stop and think about genetic counselling before they give it an unqualified seal of approval. But those people like myself who believe that, although abortion can be justified, it should not be performed without good reason, will find that diseases like haemophilia present very difficult moral dilemmas.

We come next to the position of the parents in counselling. In one sense it seems clear that inasmuch as people can avoid having genetically handicapped children their personal happiness is increased (or unhappiness decreased), and that consequently, inasmuch as such services are made generally available, generally this makes genetic counselling a good thing. There are, however, two troublesome points which must be raised.

First, in the new enthusiasm for the options opened by amniocentesis and possible abortion of the diseased, the excruciating psychological trauma of the parents involved must not be ignored. Whatever the case may be for general cases of abortion (i.e., abortion on demand for unwanted pregnancies), it seems overwhelmingly clear that couples who have a fetus aborted because it is found to be diseased suffer a great deal. One would expect this. People involved in counselling are people who want to have children — want to have the child being tested — unlike most involved in abortion. Moreover, the technical aspects of counselling imply that the fetus will be fairly well developed before a final decision can be made. First, there must be sufficient amniotic fluid for a tap to be performed (not possible before 13 weeks), and then, other than for direct chromosome studies, the cells must be cultured before tests can be made. All told, one is therefore compelled to wait until the fetus is about 17 weeks old — in short until about the time it has quickened (Bartsch, 1978). It goes almost without saying that an abortion at this stage of a pregnancy is far more traumatic than an earlier abortion, not to mention the strain of waiting until the test results are known. And what empirical evidence there is confirms the burden of the counselling experience. For instance, one (admittedly rather small) study of 13 couples who had a therapeutic abortion found that almost all involved (men and women) suffered severe depression following the abortion. Indeed, four out of the 13 couples separated during the pregnancy/abortion period (Blumberg et al., 1975; see also Fletcher, 1973).

One might argue with some justification that one has here a matter of choosing the lesser of two evils. However, one point does seem clear. An adequate counselling service is going to need comprehensive psychological back-up facilities. If nothing else, this is all going to be fairly expensive, and ought surely be acknowledged in any realistic cost-benefit analysis. It will not escape the reader that the costs of such facilities do not occur in the debit column of Kaback's cost-benefit analysis of his Tay-Sachs programme. (I shall have more to say about this analysis shortly.)

But now we come to the second major problem with genetic counselling considered from the viewpoint of the parent; the problem of coercion. The discussion so far has allowed that some abortion, specifically including that of genetically diseased fetuses, is justifiable. But what about people who, for various reasons, do not want their fetuses to be aborted, even though there is reason to believe that their children will be born genetically handicapped? What about the older woman who has tried desperately for years to have children, and now finally at 40 is pregnant with a Down's Syndrome child and wants to keep it? Or what about, say, the Catholic mother with two haemophiliac sons who is pregnant again? Can one in any sense justify compelling such a person to seek counselling and, if necessary, carrying through with an abortion of the fetus?

One's immediate reaction is to deny that one can, and what I had to say in my introduction about the foundations of moral claims, does not seem to imply the permissibility of such compulsion. Indeed, I have explicitly acknowledged that one must respect the rights of the individual, and it hardly seems that the overall happiness of the group in not now having to provide facilities for such a handicapped child, outweighs the agony of a mother forcibly aborted against her will. If the Kantian side of ethics is to have any force at all, it will come into play here. On the other hand, it is stupid to pretend that acknowledging the rights of the individual at this point makes all tension disappear. At present, we as a society accept genetically handicapped children and we try to help them and their parents: not, I recognize, as much and as generously as we should. The genetically handicapped and their relatives are nature's unfortunates, and inasmuch as we turn our backs on them we lessen our own humanity. But it is obviously the case that if the medical profession as a whole feel that they have developed a relatively efficient method of identifying and eliminating diseased fetuses, strong pressures are going to build up, insisting that these methods be universally employed. And resentments will occur when people, for whatever reasons, try to resist the pressures. "Why", it will be asked — indeed is already being asked

— "should society help support and aid genetically handicapped people, when their very existence could easily be prevented?" Already, nearly a decade ago, one discussant of the problem was led to write as follows:

Thus, while in principle the parents of a fetus with a detected case of Down's syndrome are still left free to decide whether to carry it to term, it is not difficult to discern an undercurrent in counselling literature and discussion that would classify such a decision as irresponsible. This is amplified in a subtle way. Abortion is said to be 'medically indicated' in such cases, as if what is essentially an ethical decision has now become nothing but medical. (Callahan, 1973, p. 85.)

I do not think this problem can be minimized. At the very least I suspect that genetic counselling is pointing us in a direction which will demand compulsory sterilization of those sufficiently 'irresponsible' to refuse to have affected fetuses aborted. If anyone doubts this, think for the moment how many things in our lives are now legislated although they were free a hundred years ago. It may be felt that this is an acceptable restriction on human freedom: I confess myself that I feel a lot less perturbed by compulsory sterilization than by compulsory abortion. But it must be recognized fully that this is a penalty we shall probably have to pay — at least examine — in our quest to make life better through genetic counselling.

6.5. THE PROBLEM OF THE POOR

Jesus was probably right when he said: "Ye have the poor always with you." At least, there has never been a society without its poor. If we are to pay due respect to each individual, ideally one would probably like to argue that a society's overall moral worth is a direct function of the minimum happiness attainable — that is, of the most deprived citizens (Rawls, 1971). One would, for example, give low ratings to the Third Reich because the Jews suffered so, even though life may have been pretty good for the average member of the Hitler Youth. However, as always in this imperfect world, we cannot make our criteria too sensitive. Otherwise we shall make 'overall moral worth' or 'quality of life' or whatever phrase it is that we are using to denote a good society virtually meaningless, because all societies rank equally at the bottom. Appalachian whites and urban blacks will drag the U.S. down to the level of the worst South American dictatorship.

A balance must be struck. A society should not be said to be totally deficient in moral feeling or have a low quality of life simply because it has some poor. Proportionately, however, there should not be that many poor; moreover, society as a whole should be trying to help the underpriviledged: both of this generation and the next. Nazi Germany had a low quality of life, because it was deliberately pursuing a policy of persecution against the Jews. On the other hand, a society with fewer material goods but trying to help all, might be judged to have a higher quality of life (Britain after the Second World War?).

Now, how does all of this relate to genetic counselling? Fairly directly, I suspect. In at least two ways genetic counselling represents a kind of approach to medicine which benefits disproportionately the middle classes over the poorer classes. First, it is clearly an expensive kind of medicine, and it is hard to deny that, if our concern as a society were to increase the health and happiness of as many members of society as possible, a strong case could be made for diverting funds elsewhere. Consider Tay-Sachs disease. We learnt that about 50 children per year are born so affected in the U.S. Their deaths are a tragedy, but so also are the lives of the youngsters living in city ghettos and suffering from retardation through malnutrition and other deprivations. I suspect that one could easily find 50 such cases in a city block of the average black ghetto, or, lest I sound condescendng from my position north of the border, on many of our native reservations. In fact, we know that although (in the U.S.) much retardation is caused by the genes, much more is caused by lack of good food — or any food. " . . . perhaps 3 percent of the retardation we deal with in America is genetic in origin and probably 10 percent is caused by socioeconomic deprivation. Nutritional deficiency is one of the major aspects of this" (Hilton *et al.*, 1973, p. 334). In short, what I am suggesting is that if we are really serious about raising the quality of life in North America through improved health care, it might pay us to invest in milk powder rather than genetic counselling.

Two counters will probably be made here. First, that concern in one area (bad genes), does not necessarily preclude concern in another area (bad food). Second, that even directed as it presently is to small numbers from the middle class, evidence (already given) shows that genetic counselling pays its way. It frees funds for other aspects of medical care. But neither of these counts really stand close scrutiny. Apart from anything else, one should not think of things as being simply reducible to dollars and cents. If leading and talented members of the medical fraternity are devoting their thoughts, time, and efforts to genetic counselling, there is just that much less human resource to be given to other problems. Remember also that these leaders will be influential on the rank and file — most probably they will be the teachers! And in any case, there is only so much money to go around, and if genetic counselling is getting it then nutrition is not.

However, what about the objection that cost-benefit analyses show the value of genetic counselling? Well, with all due respect, the analyses so far provided puts one in mind of Disraeli's quip that there are three kinds of untruth: lies, damn lies, and statistics. Take Kaback's (1974) figures given above, purportedly showing the cost-benefit effectiveness of his Tay-Sachs programmes. We have already seen some reason to doubt the rosiness of his figures, because he omits all mentions of the costs of psychological counselling, something which, all seem to agree, is crucially important in any genetic screening programme. I am afraid that this is one case where one will not be able to call on the voluntary services of that underpaid and overworked individual, the Catholic Chaplain. But Kaback's figures have even less connection with reality than this. In assessing the costs of his programme, he calmly omits "professional salaries or costs of volunteer efforts" (Kaback et al., 1974, p. 129). In other words, he omits himself and his colleagues from the Kennedy Institute, and the five or six hundred women who were involved in the project. Now, it might be objected that the point about volunteers is that they donate their services, and this is true — at least, it was in this case. But this does not deny that the professionals have to be paid, and their costs must surely be reflected in any adequate analysis. Moreover, it should be remembered that the Tay-Sachs programme was peculiar in that it was confined to Jews. Their community contains financially comfortable, intelligent, educated people. In other words, their's is the kind of community where one might hope to get a high number of unpaid volunteers — particularly since Tay-Sachs is so distinctively a problem for Jews and not for others. In short, one cannot assume that for genetic counselling generally one can draw on so many unpaid volunteers, who were apparently able to do quite important tasks (e.g., recording, tube handling, and labelling, (Kaback et al., 1974, p. 110)) — not to mention the twenty-odd volunteer physicians at each screening session. A realistic cost-benefit analysis, one which is truly generalizable, must reflect the costs, not only of the physicians, but of the other needed personnel also.

I remain unconvinced, therefore, that genetic counselling is such a profitable form of medicine, and that because of its savings more resources will be freed for the poor. (I have similar reservations about other cost-benefit analyses, like Mikkelsen et al. (1978), Glass and Cove (1978), Hagard and Carter (1976), Hagard et al. (1976).) But there is a second reason other than financial why genetic counselling might favour the middle classes disproportionately. Having the potential for a genetic disease is not like having a ruptured appendix. With the latter, one has a pain in the belly and knows only too well that one needs help. The former, even though the effects might be just as devastating as the appendix, needs trained intelligence and knowledge to appreciate, particularly when carriers are themselves perfectly fit. In other words, genetic counselling, either on an individual basis or through mass screening programmes, is likely to be more appealing to and therefore more utilized by those further up the socio-economic scale.

The extent to which this kind of expectation is realized is illustrated dramatically by Kaback's Tay-Sachs programme. What I have said before about the nature of Jews is strongly confirmed. Some 7000 people were tested in the first year of the programme. "Approximately, 75% percent of those tested were either graduates of college or college and post-graduate programs. Less than 1 percent had not completed a high school education" (Kaback *et al.*, 1974, p. 119). And a follow-up study, comparing those that came for testing, with those that did not, showed significant differences ($p < 0.05$).

Respondents were significantly younger, had fewer children, were less likely to have completed their families, were more educated, and were higher on the social index scale. In addition, and perhaps most importantly, by several criteria, the participants appeared to understand more clearly the genetic information regarding carrier risk and implications and the subsequent alternatives if they alone or with their spouses were found to be heterozygotes. (Kaback *et al.*, 1974, p. 127.)

As one might expect, others have found that it is invariably the better educated, more successful, who take advantage of genetic counselling facilities. Just such a similar story can be told from Canadian experiences of amniocentesis. Of some 1000 cases reported in one study (in 1976), involving tests for all kinds of disease, and potentially involving people from all sequents of society (i.e., not just Jews) a far higher proportion than normal had upper-level education backgrounds. "Twenty-seven percent of the women who underwent amniocentesis had completed a university education, while 19% had completed only grade 8; the comparable figures for the general Canadian female population were 7.5% and 44%. The head of the household had a professional, technical or inaugural position almost twice as often in the amniocentesis group as in the General Canadian population" (Simpson *et al.*, 1976, p. 740). Interestingly, although expectedly there were proportionately more Protestants than Catholics than in the general population, the absolute numbers of Catholics was quite high. (Amniocentesis group, Protestant to

Catholic, 50% : 30%; general population, 37% : 46%). Clearly some Catholics, their Church's stand on abortion notwithstanding, will use genetic counselling facilities. Of course, this does not deny that many will not, as the Canadian figures bear out. (See also Galjaard, 1978; Passarge, 1978; Brock, 1978.)

With reason, it might be objected that one ought not to deprive one segment of society (the middle class) from modern medical advances because another segment (the poor) refuses to take advantage of them. After all, I enjoy watching PBS (Public Television), even though most people shun it like the plague. But it seems undeniable that inasmuch as one promotes genetic counselling programmes to improve the overall quality of life, to make a better society, one ought at the very least take special precautions not to exclude the poor (specific advertising programmes?), and that inasmuch as one aims to improve health care generally, one must weigh carefully costs and benefits of new medical technology as opposed to more conventional and less exciting approaches. Certainly one must weigh things more carefully than have been done to this point. As a start, I would suggest that the doctors get some help with their sums. I am all too aware that medical people think that they and only they can solve problems of medical ethics; apparently, they feel the same about medical economics too.

6.6. THE PROBLEM OF MINORITIES

Minority groups are not necessarily poor or underprivileged, although they often are. Hence, much of what has just been said about the implications of the kind of medicine represented by genetic counselling will be pertinent here. However, genetic counselling poses problems for some people specifically because they are parts of minorities rather than because they are poor — even though they may very well be poor too. This is because certain genetic diseases are very closely identified with certain minority groups, rather than with the population as a whole: obvious examples being Tay-Sachs disease in Jews and sickle-cell anaemia in blacks. Hence, unless one treads very carefully, in the quest to raise the overall worth of a society one runs the grave risk of providing yet more examples of the majority imposing its 'norms' on minority groups, whether or not the minority groups want to share in these norms. In other words, utilitarian urges may violate Kantian barriers.

The dangers I am hinting at do not necessarily obtain. Apparently the Jews of the Washington-Baltimore area participated willingly in the Tay-Sachs screening programme. Indeed, the programme could not have been the success that it was without the active cooperation of both rabbis and lay-people. And certainly, no one seems to have looked upon the programme as a subtle form of anti-semitism by members of the Kennedy Institute. On the other hand, we know that this programme was peculiar in that the Jews are well educated — individuals could themselves see the dangers and make their own choices — and, of course, Tay-Sachs is an unambiguously horrible disease. But if one is dealing with a group which is poor, ill educated, and which has reason to distrust the majority — like blacks and native peoples — and if also the disease is not quite so dreadful, the potential for misunderstandings and ill-will rises dramatically. As also does the potential for minority-oppressings biases by genetic counsellors: biases, I readily admit, which may be subtlely concealed beneath layers of paternalistic (or maternalistic) altruism.

I am not saying that nothing can be done about the problem of genetic disease and minorities. I am not saying, for example, that no screening should be done for sickle-cell anaemia in North American blacks, although I do suspect that the only effective way this case be tackled is with major motivation from the blacks themselves. What I am saying is that the whole question of minorities and diseases peculiar to them must be of concern to genetic counsellors precisely because these are minorities involved. I would even go as far as to say that I do not find it inconceivable that one might have some not-too-dreadfully-bad ailment, and that were it in the general population one might try to do something about it, but because it is confined exclusively to some oppressed group, one might decide to do little: paradoxically, not because one wanted to go on oppressing the group, but because such a 'cure' might too easily be taken to be worse than the disease.

It may be felt at this point that I am being overly pessimistic about genetic disease and minorities. Let me point out that the one history of a wide-spread disease in a sizeable minority, namely sickle-cell anaemia in blacks, does not give cause for optimism (Reilly, 1975; Erbe, 1975; GSR, 1975). One has a trial of hasty and ineffective legislation, not to mention genuine injustice from a succession of half truths, by no means maliciously disseminated. It was, for example, widely and mistakenly argued that sickle-cell trait (i.e., possession of one sickle-cell gene heterozygously) virtually reduces the bearer to an invalid's life — this was certainly a rumour effective enough to increase significantly the insurance premiums of bearers. Perhaps one might feel that these are problems that anyone is liable to encounter as we get to grips with genetic disease: I defy anyone to deny that all of history cries out that they will fall more heavily on minorities.

I might add parenthetically that, whereas until recently, only heterozygotes for sickle-cell anaemia could be identified, now it is possible to identify in utero fetuses homozygous for the sickle-cell gene (Kan *et al.*, 1976; Valenti, 1978). And already, the developers of the technique are priding themselves on the possibility of doing for sickle-cell anaemia what is done for Tay-Sachs disease: "We shall ultimately have the opportunity to approach this class of disorders with similar mass screening efforts" (Epstein and Golbus, 1977, p. 710).

6.7. WHAT IS GENETIC DISEASE?

In trying to relate genetic counselling to its moral worth, to the extent that it might be said to elevate a society's quality of life, I have assumed deliberately that the notion of 'genetic disease' is relatively clear cut. But obviously it is not, and no discussion of genetic counselling and the quality of life would be complete without some acknowledgement of this fact. In the final essay of this collection I shall be looking at some of the recent attempts to define such difficult notions as 'health', 'disease', and 'illness', and the interested reader might care to look ahead and see how the various suggestions might be applied to defining the specific notion of genetic disease or ailment. But it is not really necessary here to get into a formal analysis — a few fairly obvious examples can make the points about the difficulties that will be posed by the need to make decisions about what counts as 'disease' and more particularly what counts as 'severe disease' (i.e., one sufficiently severe to warrant sustained effort to detect and eliminate).

Presumably, if one endorsed genetic counselling at all, by whatever criterion one used, some diseases would be caught and considered sufficiently severe to warrant screening, amniocentesis, and abortion. Tay-Sachs disease is an obvious example, and there are some which are even more horrible: Lesch-Nyhan's syndrome is marked by mental retardation and compulsive self-mutilation. But what about something which is less devastating? Diabetes is undoubtedly a disease, and certain forms of it have a genetic basis (Levitan and Montagu, 1977). But is it sufficiently severe to advocate genetic counselling? Perhaps one might decide that it is something of a border-line case. Thanks to insulin, diabetics today can live full and useful lives, but not without inconvenience and the ever-present threat that something might go wrong and cause severe problems, like blindness. Hence, one might want to argue that no general screening programmes for diabetes should be set up, but that services be made available for individual couples who want counselling.

Nevertheless, as one descends the scale, problems about that constitutes a genetic disease intensify, as do related problems about when a genetic characteristic is of a type that counselling – either state endorsed or individually sought – is appropriate. And indeed, given the condition of our society, the potential seems ripe for significant social and psychological disruption. Take, for example, homosexuality. One might or might not think that homosexuality is a disease. This is a matter which we shall discuss later (Essay 10). But whatever we may decide, there is no doubt that many do think it a disease, if not something very much worse (Bieber *et al.*, 1962). Now, there is some suspicion that some forms of homosexuality orientation may have a genetic foundation, although as yet no one seems to have much ideas about the precise etiology (see Essay 8). Suppose, however, that the pertinent genes were identified, and that, in practice, it became possible to screen for them. One would be a fool to deny that such a power would have the most severe consequences: particularly in light of the fact that the evidence is overwhelming that today few subjects can be more calculated than homosexuality to set children against parents and factions against factions. We have come a long way, but homosexuals are still persecuted and discriminated against (Knutson, 1980). If there existed the possibility of eliminating (some) homosexuals before birth, there would be great pressure by many to do so. And this activity, value-laden in its hostility to homosexuality, would naturally lead to increased insecurity by those homosexuals who are born – not to mention an overall decrease in tolerance of homosexuality and desire to understand its implications.

Or finally consider the ultimate genetic 'affliction', sex. Apparently most people who want children want children of both sexes, but they want boys first (Westhoff and Rindfuss, 1974). Are we on the verge of allowing people to have abortions simply because their children will not be of the sex that they want? I must add that genetic counsellors are not advocating or allowing this, even though the required technology is already at hand. (The exception is for sex-linked diseases like haemophilia where people are counselled to have all male children aborted.) On the other hand, what is possible today has a nasty habit of becoming permissible or even obligatory tomorrow. If anyone doubts this, consider the present pressures on unmarried pregnant teenagers to have abortions. Already, at least one knowledgeable thinker in the field has reversed himself, now agreeing that counselling for sex selection ought to be allowed (Fletcher, 1979). And yet, if sex determination is generally possible through counselling, I foresee psychological and social problems as grave as those from an attempt to stamp out genetic homosexuality. How would you, a girl, feel if you knew that your parents had aborted a couple of girls before your older brother was born? For that matter, how would you as the older brother feel? Is it entirely ludicrous to wonder whether, if parents did generally choose to have a boy first, this would lead to further ingraining of sexism in our society — if only because older children tend to be dominant achievers and, therefore, if older children are generally boys and girls are generally younger children, then psychological pressures will reinforce the imbalance between the sexes which we already have today?

In summary, what the few examples given in this section seem to show is that there is no firm, unequivocal line to be drawn between genetic afflictions universally agreed to be so severe as to merit counselling and those that are not. Furthermore, it seems plausible to assume that as counselling gets more familiar and more sophisticated, its range is going to widen to such a point that severe social and psychological turmoils may follow in its wake.

6.8. CONCLUSION

By now the reader will have realized that I am somewhat schizophrenic about the moral value of genetic counselling. On the one hand, it is undeniable that genetic counselling can avert or soften human tragedy. This in itself is surely a potential to raise the quality of life, and leads us to judge genetic counselling as a good thing. Moreover, a point which has not been emphasized but which is pertinent: genetic counselling is obviously a very exciting field for the medical researcher. Those who believe, as I most certainly do, that there is more to life than mere existence, will agree that inasmuch as a society allows and encourages its brightest members to press forward the boundaries of knowledge, the society is that much more worthwhile. And this holds whether the boundaries are medical technology or philosophy! Hence, for this reason also genetic counselling counts as a good thing. (In this context, see also the next essay.)

But, on the other hand, the arguments just presented surely show that genetic counselling, particularly as it gets more widespread, will not unproblematically raise the quality of life of any society, leading one to endorse such counselling unequivocally. Indeed, genetic counselling has the potential, already verging on the actuality, of causing great social harm. It can and very likely will cause severe tensions for people taken both individually and in groups. So, what can be done about my schizophrenia, other than pointing out that this is an afflection with a genetic base and no doubt someday will be the subject of a screening programme? I do not call for an absolute moratorium: but there is obviously a desperate need for detailed studies of genetic counselling, as it is today and what it implies for tomorrow. At the moment we seem to be on a speeding train driven solely by medical technologists, who think that because they have developed and administer the technology, this qualifies them to talk about and decide on all the related topics, whether these be philosophical, economic, or whatever. But if nothing else, this discussion should show that this conclusion does not follow. Of course, medical technologists must have a special place in the debate, but not the only one, particularly if society at large has to pay most of the costs. Genetic counselling is a human opportunity and challenge. It has the potential to raise the quality of our lives, and to depress them. It would be wrong to let its power and direction rest, as today, solely in the hands of enthusiastic practitioners.

Does genetic counselling raise the quality of life? Possibly!

NOTES

[1] Ashkenazi Jews are Jews from Eastern Europe, as opposed to the Sephardic Jews from Spain and Portugal. In North America, 9 in 10 Jews are Ashkenazi.

[2] In case it be objected that it is improper to compare living diseased people with aborted fetuses, healthy or otherwise, I might note that there is some evidence that amniocentesis causes respiratory problems and 'major orthopaedic postural abnormalities' in 1–1.5% of infants (MRC, 1978, p. 1).

[3] In this essay I shall not discuss explicitly the morality of abortion, not because it is an easy matter to decide, but precisely because it raises so many problems, and would therefore be liable to swamp my whole essay. Side-stepping the main issue I assume without argument what is apparently a somewhat conservative stance: abortion can be justified, but should not simply be regarded as a form of contraception (see Feinberg, 1973).

[4] The genetics of this situation can be explained easily. We know that human beings are sexually different because, whereas females have two so-called 'X' chromosomes, males have one 'X' chromosome and one 'Y' chromosome (i.e., females are homozygotes and male are heterozygotes). The X chromosome is much larger than the Y chromosome and, therefore, carries more genes. Suppose that on the X chromosome (but not the Y) there is a gene B, recessive to A. We have, in reproduction, the following Mendelian situation.

	MALE	FEMALE
	Y| |X	X| |X
Parents	⊥ A	A⊥ ⊥B
	(A caused phenotype)	(A caused phenotype)

Offspring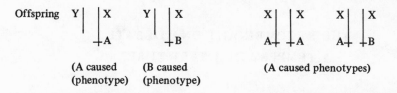

(A caused   (B caused   (A caused phenotypes)
(phenotype)   (phenotype)

If B is the haemophilia gene, it can be seen that only sons are physically sick, other sons do not have the gene at all, and half the daughters carry the gene.

THE RECOMBINANT DNA DEBATE:
A TEMPEST IN A TEST TUBE?

This world is given to us on loan. We come and we go; and after a time we leave earth and air and water to others who come after us. My generation, or perhaps the one preceding mine, has been the first to engage, under the leadership of the exact sciences, in a destructive colonial warfare against nature. The future will curse us for it. (Chargaff, 1976, p. 940.)

Ten years ago, if a biologist had read this passage, the final paragraph of a letter to *Science*, the rather smug assumption would have been that 'obviously' the author was writing about the threat from atom bombs or nuclear reactors or chemical pollutants or some such things: nothing to do with biology. But this is not ten years ago, and the letter, written in 1976 by Erwin Chargaff, Professor Emeritus of Biochemistry at Columbia University, was indeed about biology and the threat which he saw it posing to humankind. What Chargaff was decrying specifically was a powerful new biological technique, one which enables scientists to extract the DNA of one organism and insert it into the DNA of another organism, thus giving rise to all sorts of hybrids apparently not existing naturally. He argued that we have no way of forecasting what this 'Recombinant DNA' might do, and that for all we know otherwise it could cause irreparable harm to us and further generations. Hence, all research of this nature should be stopped, or at the very least carried on in an extremely limited fashion under the most stringent of conditions. "You can stop splitting the atom; you can stop visiting the moon; you can stop using aerosols; you may even decide not to kill entire populations by the use of a few bombs. But you cannot recall a new form of life. An irreversible attack on the biosphere is something so unheard of, so unthinkable to previous generations, that I could only wish that mine had not been guilty of it" (ibid.).

When one finds scientists say this sort of thing, a philosopher's mouth starts to water. Surely to goodness there must be some nice juicy issues which take us beyond 'pure' science, whatever that might mean, to matters to do with the whole nature of science, its values and responsibilities: the very things, in short, that philosophers care about. Let us, therefore, take a look at the 'recombinant DNA debate' seeing if some factors do indeed emerge which merit an analytic examination. (A good history is Grobstein (1979); see also Richards (1978a), Jackson and Stich (1979).)

158

7.1. THE RECOMBINANT DNA DEBATE

The recombinant DNA debate began at the beginning of the decade. As molecular biologists developed and mastered ways of combining DNA from different organisms into functioning units, a number of them — by no means all as shrill as Chargaff — began to wonder and worry about possible dangers. These fears were not without foundation. One proposal, for instance, was to take DNA from an organism that causes cancer-like growths and to transplant it to another organism, the bacterium *Escherichia coli*: a bacterium which lives quite happily in the bowels of every one of us. The prospect of such a newly created life form getting loose and doing more damage than all the tobacco in Virginia (or Ontario) is, surely, one calculated to make even the more headstrong amongst us pause and think again. Thus, concern grew until, at an international meeting of the biological researchers themselves, in Asilomar, California, in February 1975, it was agreed to try to regulate recombinant DNA research, avoiding the most potentially hazardous experiments and keeping all work within certain common-sense safety standards (Berg *et al.*, 1975).

Although the scientists had not then played Dr Frankenstein, creating a monster from many different organisms which ran wildly out of control, they did apparently create something they could not control. The very fact that they felt they needed guidelines and safety rules at all, created and alerted critics, both within and without the biological community. With increasing degrees of vigour it was argued that no guidelines can possibly be adequate, there are bound to be mistakes which endanger us all and, thus, many critics called for a total ban on recombinant DNA research or, at least, for very restricted work. Perhaps not surprisingly, the call for curtailment was even heard from outside the university community. Cambridge (Massachusetts) City Council, for instance, tried to stop such recombinant DNA research at Harvard and MIT (Grobstein, 1979, chapter 5; Zander, 1979).

Thus, matters escalated until 1977, when, I think it is true to say, the whole debate began to subside as quickly — if not more quickly — than it had blown up. Stiffer and more effective regulations somehow never got passed into law (I speak now specifically about the U.S., but the story was not too different elsewhere), and then even those rules which had been put up got watered right down (Wade, 1978a, 1979a). Scientific opinion swung to the view that the dangers of recombinant DNA work had been vastly overestimated. Today in fact, much recombinant DNA work is allowed to go on virtually anywhere, so long as one closes the door when one is working and puts on a lab coat if the head of the team insists (DHEW, 1980). As all authors know, book manuscripts tend to take a mysteriously long time to appear in print: thus only now are full-length discussions of recombinant DNA, and the debate it caused, appearing. Knowledgeable reviewers moan that everything is dated and no longer relevant (Davis, 1979b; Krimsky, 1979).

It might seem, therefore, that the proper person to look at the recombinant DNA debate is the historian of science not the philosopher of science. If indeed recombinant DNA work holds no danger, why bother to talk about the morality or whatever of allowing it or not allowing it? Nevertheless, I think there are at least three reasons why a philosopher might fruitfully take a retrospective look at recombinant DNA work. First, with all due respect, I am simply not that trusting of scientists. Scientists like the rewards that their profession has to offer: Nobel prizes, jobs at Harvard, even tenure at Hicksville State University, not to mention the glow of pride which comes from looking at long columns after one's name in the *Science Citation Index* (Cole and Cole, 1973). Now, it seems to be agreed by all that recombinant DNA opens up an incredibly powerful new way of studying the biology of the cell (Jackson, 1979; Kushner, 1978). One can find things out more quickly and in much greater detail than one can by other methods. Hence, to the scientist a fairly strong element of self-interest enters into the decision about whether or not to allow recombinant DNA work. Consequently, I think at the least that the rest of us are entitled to see if the present optimism about the safety of recombinant DNA work is justified. I might add that our right and need to look into this matter is compounded by the fact that there are still some critics within the scientific community who question the general optimism (Rosenberg and Simon, 1979). Moreover, the truth that accidents can happen is underlined by the fact that recently a very unpleasant organism, the virus causing foot and mouth disease, escaped from one of the U.S. Government's most carefully restricted facilities (Wade, 1978b). As pointed out, today much recombinant DNA work is done under virtually no restrictions at all, and even as I write there has been controversy about an organism which was or was not what it was supposed to be, which was or was not cloned, and which was or was not on a restricted list (Wade, 1980). Thus, whilst I would agree that recombinant DNA researchers themselves should and must have input into decisions about the permissibility of their work, I think it would be wrong if the rest of us were not to have an input also.

My second reason for concern and for feeling that the recombinant DNA question should not be closed too quickly, is that recombinant DNA research is rapidly becoming a very big business, with large individual and corporate fortunes riding on it (Wade, 1979b, c). One does not have to be a Marxist to acknowledge that unenlightened self-interest frequently enters into decisions where big money is concerned. Frankly, I feel a little uncomfortable about anything which attracts the attention of such corporate giants as Dupont and Standard Oil; and they are both investing heavily in recombinant DNA work. I am not arguing that investing in science hoping to make profits is bad *per se*. I do argue that the rest of us should reserve the right to look at what is going on.

Third, I think that some interesting general philosophical points come out of the recombinant DNA debate. Whether they will be needed again for resolving problems arising out of recombinant DNA work, perhaps remains to be seen. But undoubtedly the tension between science and the rest of life will flair up again, and so the more we can learn now, the further ahead we will be the next time. (See also Callahan (1978), Richards (1978a, b), Williams (1978), Stich (1979), Cohen (1979).)

Enough of the reasons why a philosophical outsider might continue to look into recombinant DNA research. Let us move on next to specific questions which must be answered before overall conclusions can be drawn: What does recombinant DNA work involve? What does it promise? What does it threaten?

7.2. THE NATURE OF RECOMBINANT DNA RESEARCH

The reader will be getting tired my constant refrain, but like almost everything else in molecular biology, the basic ideas behind recombinant DNA research are really quite simple.[1] As always in molecular biology, we start with the double helix. This entwined pair of macro-molecules is the carrier of heredity and the unit of function for the widest range of organisms under the sun. Moreover, single DNA strands consist invariably of combinations of the four basic nucleotides, linked one after another (see Essay 5). In theory therefore, there is no reason why a piece of strand from one organism should not be linked up with a piece of strand from another, widely different organism, say like a duck with an orange. To use a metaphor, it is as if one were trying to link up two pieces of string, of exactly the same material and of exactly the same diameter. It is not as if one were trying to join a piece of string with a piece of rope, or even worse, a piece of string with a piece of chain. The problem posed by breaking up and recombining DNA segments, therefore, are more practical than theoretical. How can we get pieces of DNA, perhaps from quite different organisms, to join up? And how then can we get

the newly created hybrid DNA pieces to start working, and produce something in the way of proteins and so forth?

There are several methods that can be used for the actual recombining
(Jackson 1979). They all rely on the fact that in the double helix invariably
we get one kind of nucleotide on one DNA molecule pairing with another
fixed kind of nucleotide on the other molecule: guanine (G) pairs with
cytosine (C), and adenine (A) with thymine (T). The trick is to produce a
helix with one or a few spare nucleotides on the end of one of its strands,
which can then be used to link up with another helix, also with one of the
strands having spare nucleotides, and where the spare nucleotides on the two
helices are *complementary* to each other. The two helices will then fit neatly
together and the actual DNA strands can be bonded without any break at all.
The most straight-forward, although technically rather complex way of doing
all of this is as follows. We take a sequence of DNA helix and add 'tails' to
the ends of the DNA molecules with the aid of an enzyme ('terminal trans-
ferase'). We can control the nature of these tails by restricting the availability
of the bases and, thus, they can in fact be made of identical bases. Simultaneously we add tails to another sequence of DNA, making sure that the
bases are complementary to those of the first sequence. Then the sequences
of DNA are mixed, thanks to their 'sticky' ends they join up, and then another enzyme ('DNA ligase') is used to make the joins permanent (Figures 7.1
and 7.2).

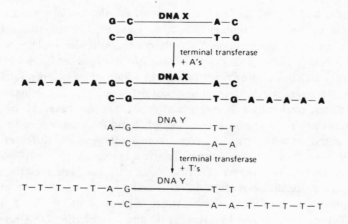

Fig. 7.1.   Adding complementary tails to DNA (from Jackson and Stich, 1979, p. 44).

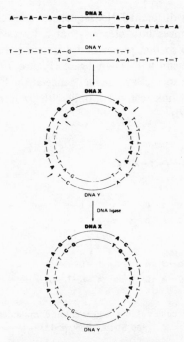

Fig. 7.2. Making a recombinant DNA molecule with tails formed as shown in Figure 7.1
(from Jackson and Stich, 1979, p. 46).

An alternative and much-favoured method today uses a set of enzymes
known as 'restriction' enzymes. These enzymes cut DNA helices unevenly,
but they always cut in the same way. One thus always gets certain tails of
specific sequence (Figure 7.3). The recombining procedure therefore is quite

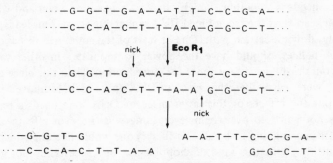

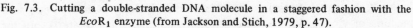

Fig. 7.3. Cutting a double-stranded DNA molecule in a staggered fashion with the
$EcoR_1$ enzyme (from Jackson and Stich, 1979, p. 47).

simple. One mixes up different DNA sequences, adds the restriction enzyme which cuts the sequences into pieces — but always with the same kinds of tails — stirs things up so that pieces find different partners, and then 'glues' everything together again with DNA ligase (Figure 7.4). The whole phenomenon is very similar to the 'excuse-me' dance at the high school prom, even down to the fact that links can occur only with complements (i.e., C with G, A with T, boy with girl).

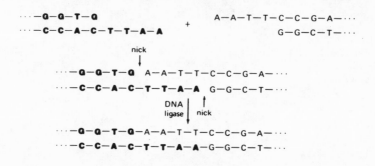

Fig. 7.4. Joining up molecules cut in a staggered but complementary way (from Jackson and Stich, 1979, p. 48).

Getting everything to work, once it is put back together again, requires more ingenuity. To continued the colourful illustration of one of the leading researchers, David Jackson (1979), suppose one combined the genes of a duck and of an orange. Do not think that one can sit back and that shortly out of the test-tube will come the tooth-some smell of duck *à l'orange*! If circumstances are not right, nothing at all will happen: the DNA will just stay put. What one needs is something which can start replicating again and which, in particular, can start the recombined DNA replicating again. This leads, almost inevitably, down a certain path. One at least of the combined pieces of DNA (or DNA helices) should have the power to replicate: in other words, it should contain the necessary genes for reproduction. Then this piece of DNA can, as it were, carry the other piece of DNA on its back. Moreover, if one is to sort out the effects of this latter piece of DNA, one needs a replicator which is not going to overwhelm everything with its own effects, and obviously (if one is going to be all this fussy!) one would like one which reproduces quickly. With this kind of shopping-list, one is certainly not going to look at elephants! Rather, one finds oneself looking at micro-organisms: bacteria or viruses. Viruses have certainly captured some interests, but the most popular organism for recombinant DNA work is the well-studied *E. coli*. However, it is not the main DNA strands in this organism which get researchers' attention: work concentrates rather on small autonomous circular pieces of DNA within the cells: 'plasmids'. Through various subtle techniques, one can extract the plasmids from *E. coli* cells, insert fragments of DNA from other organisms into them, and then put the plasmids back in *E. coli* cells, where they can again be 'switched on'. As they replicate the foreign DNA replicates also, and thus one can study the special effects, if any, of this newly introduced genetic material (Figures 7.5 and 7.6).

But where and how does one get the foreign material to put into plasmids, viruses, and the like? Sometimes, although the choice of originating organism is controlled, there is a little or no selection over which pieces of foreign DNA are introduced into a replicating unit. In so-called 'shot-gun' experiments, a large part of the entire DNA of an organism is broken into small parts by

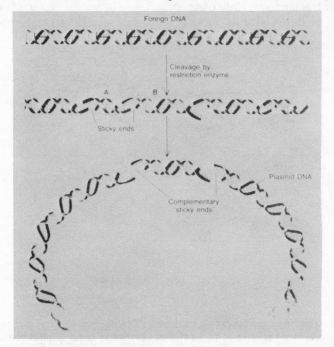

Fig. 7.5. Recombinant DNA technique. A piece of DNA (that at the top, labelled 'foreign') is cut up by a restriction enzyme, which same enzyme is used to cut open a plasmid. Then the foreign piece is inserted in the open plasmid, and everything is joined back up again (from Grobstein, 1979, p. 24).

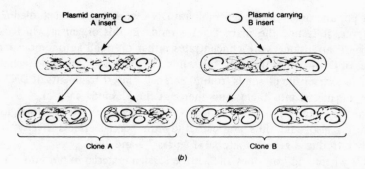

Fig. 7.6. Here, two plasmids carrying foreign DNA 'inserts' (as shown in Figure 7.5) are returned to *E. Coli* cells, where they propagate, thus forming two 'clones'.

appropriate restriction enzymes. Then these fragmented parts are inserted, more or less randomly, into plasmids and thus into most bacteria (Figure 7.7). Clearly, in theory at least, more or less anything might get taken up from the

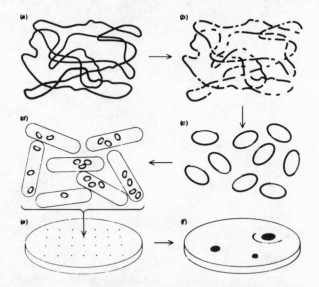

Fig. 7.7. The Stages in a Shotgun Experiment. "In a 'shotgun' experiment, the total DNA of an organism (a) is exposed to restriction enzymes in order to yield many fragments (b), which are then recombined with the DNA from a suitable vector (c) and randomly reinserted with the vector into the host cells (d). The *E. coli* hosts are next to spread on a nutrient substrate (e), so that each recipient cell, containing a particular inserted foreign nucleotide sequence can grow into a colony (f)" (from Grobstein, 1977, p. 43).

fragmented DNA, and incorporated and replicated by the bacteria. There are, however, ways of exercising more control over the DNA to be inserted into replicating organisms or their parts. For instance, under some circumstances, special techniques enable one to get fairly large quantities of RNA (which has been copied off DNA). Then, with the aid of a special enzyme, a 'reverse transcriptase', one can work backwards to get larger quantities than normal of some specific DNA sequence (Figure 7.8). This identified DNA can then be introduced into replicating units, and thus large quantities can be produced — virtually as much as one would like.

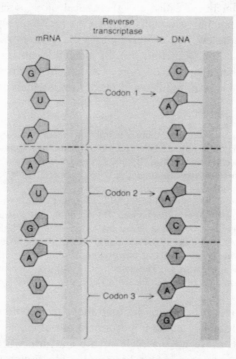

Fig. 7.8. Here DNA is being produced by copying backwards from mRNA (messenger RNA). Remember 'codons' are units of three nucleotides that code for a particular amino acid (from Grobstein, 1979, p. 57).

7.3. THE POSITIVE CASE FOR RECOMBINANT DNA RESEARCH

As I have said, in theory everything in recombinant DNA research is really very simple; in practice of course, the steps required are frequently difficult and sophisticated, taxing theoretical and technical ingenuity to the limit. But let us move on now and look at reasons why recombinant DNA has caused so much excitement. First, there is the positive case to be made. Why do molecular biologists think that recombinant DNA work constitutes so great an advance? There are two sides to the answer to this question, or at least I should say that it is generally assumed that there are two sides to the answer to this question. Later I shall be looking at a critic who denies the division, but for the moment I shall assume it without question.

First, in reply to questions about the virtues of recombinant DNA research one has what might be called the theoretical or pure-scientific answer. As hinted at earlier, the recombinant DNA methods supply one with an extremely powerful new way of delving into the mysteries of the cell. Because of such methods, one can find out an incredible amount about the nature and causes of heredity and organic development, both in simple organisms like bacteria and also in far higher forms of life, reaching right up to mammals and to humans.

With extremely rare exceptions, there is simply no practical way except by using recombinant DNA methodology to study the genes of higher organisms at the molecular level. If we wished to isolate a milligram (0.000035 ounce) of a particular gene from a mouse, we would have to start with the total DNA from about 300 000 adult mice. We would somehow have to identify and purify this one milligram of the one gene we wanted from the approximately ten pounds of DNA we would extract from these mice, and we would have to do it in 100 percent yield. This is simply not possible now or in the foreseeable future. On the other hand, if we had our mouse gene linked to an appropriate plasmid vector, we could easily produce a milligram of it, in a form that could be readily purified, by growing a few quarts of bacteria containing the recombinant plasmid overnight at a cost of a dollar or so. The actual purification of the gene would take another two or three person-days of work. (Jackson, 1979, p. 53.)

It must be noted that there are limitations to recombinant DNA research and some of these will prove important to our discussion later. Particularly as one starts to move up the scale of life and deal with higher organisms, difficulties start to mount. One has trouble identifying the genes of higher organisms, even after one has got them into bacteria; one has trouble making them work in bacteria (one has trouble making them switch on and produce polypetide chains); and if one tries to combine gene bits with functioning elements of cells much higher than bacteria, then one has all sorts of problems with understanding and controlling the whole operation: "So, although it is quite easy to isolate and amplify random bits of DNA from the genomes of higher organisms, it is still very difficult either (a) to identify what genes are carried on one of these randomly selected bits or (b) to select only those recombinant DNA molecules that contain some specific gene" (ibid., p. 52). These are difficulties and challenges, not absolute barriers.

The other side of the answer to the question about the value of recombinant DNA work concerns the potential practical applications of such research. These days when scientists look for large sums of money to fund their research, whether it be from the government or from industry, they find it politic to find or think up possible beneficial technical consequences of their work. Not entirely facetiously it has been suggested that the scientist who discovers the cause of cancer will get little praise from his fellows, who will then loose their most lucrative source of research funds. Expectedly, recombinant DNA workers have painted glowing pictures of the benefits to humankind that their efforts will bring (R. C. Valentine, 1978; Chakrabarty, 1979). To hear some of them, the pot of gold lies not at the end of the rainbow but at the end of the double helix. For them perhaps it does (Wade, 1979b).

Most obviously, there are the putative medical benefits from recombinant DNA work. The cause of many human diseases, genetic or otherwise, lies ultimately in the fact that the sufferers' bodies cannot synthesize certain essential substances: substances which healthy people can create. Perhaps best known of these diseases is diabetes, which is caused by the body's inability to synthesize insulin. It is true that today we can often obtain required substances from other sources, and then can administer such substances therapeutically to those who need them. For instance, insulin can be obtained from pigs and cows, and then given to sufferers from diabetes. However, often extraction processes are difficult, expensive, and time consuming, not to mention the fact that it is not always the case that a substance performing a certain role in an animal is sufficiently close to its human equivalent to work adequately in humans. One much-touted prospect for recombinant DNA work, therefore, is artificial production of human body products, which could then be given to those in need. For instance, if one could isolate the human gene for insulin manufacture and put it into a micro-organism, one could then hopefully produce with little bother and even less expense, as much human insulin as one liked or needed (Grobstein, 1979, p. 55).

Taking matters one stage further, it has even been suggested that one might get the body itself to help in the manufacture of required substances. Suppose one has a sufferer from sickle-cell anaemia. This is know to be caused by a specific mutation of a specific gene. (Refer back to Essay 5.) Hopefully one might put healthy specimens of this gene into viruses; then the viruses could be made to infect cells of the sickle-cell anaemia sufferer; and finally these infected/supplemented cells could be returned to the sickle-cell anaemic. With luck — or rather, as a result of anything but luck — such an individual might now him/herself produce the body-substances essential for a healthy life (Chakrabarty, 1979, p. 65; for examples of pertinent work in this direction see Rabbitts, 1976; Proudfoot, 1980). This may all sound a bit Lamarckian, but there is no question of the thus-helped anaemic being able to pass on healthy genes to his/her offspring — except if his /her sex cells got infected with the introduced virus.

Looking farther afield beyond medical prospects, hopes have also been expressed that it might prove possible to use recombinant DNA in agriculture (R. C. Valentine, 1978). One project aims to reduce the farmer's dependence on artificial fertilizers: substances which are frequently made from ever-scarcer, ever-more-expensive, foreign petroleum products. A major reason why crops like wheat and corn need fertilizers is that they require nitrogen for growth and that they cannot synthesize useful products containing it for themselves. Hence, they have to have it supplied. However, nitrogen is an abundant element — it is indeed 80% of the air — and there are micro-organisms which can take the free nitrogen and use it to their own ends. Some researchers have, therefore, suggested that recombinant DNA techniques might be used to incorporate into plants, or at least into organisms symbiotic with the plants, genes or carriers of genes which can produce their own nitrogen products. Thus, there would be little, or at least much-reduced, need for fertilizers. Intensive work is, in fact, being directed towards this end at this time.

Finally, in this listing of hypothesized technological benefits of recombinant DNA research, let mention be made of a newly created organism by Ananda Chakrabarty (Wald, 1976). This organism combines plasmids from several strains of *Pseudomonas*, each of which strains can digest crude oil partially. Reportedly the new organism is capable of swallowing such oil whole, and thus hopefully will do much to reduce the after-effects of gross oil-spills at sea. Already the legal implications of Charkrabarty's work have been discussed by no less a body than the U.S. Supreme Court, and patents have been granted on this 'new' form of life (Wade, 1979c).

Totalling things up, we see that the promised benefits of recombinant DNA are great, although it must be emphasized that to date (1980) no one has brought any of these projects to full and profitable conclusion. But the time has come now to look at the other side of the coin: the supposed dangers and limitations to recombinant DNA research and its products. At this point

I shall deal with more concrete matters, leaving more philosophical queries until later.

7.4. THE NEGATIVE CASE AGAINST RECOMBINANT DNA RESEARCH

We know that the leading and obviously most important objection to recombinant DNA work, that which caused all the controversy in the first place, is that centering on the danger. Suppose, intentionally or unintentionally, one introduced into a bacterium or a virus some gene with lethal effects on humans. Could one hope to prevent such a micro-organism from ever getting out, particularly if one allowed recombinant DNA research to go on all over the campuses of the major and not-so-major universities of the world? Moreover, it is almost as if recombinant DNA researches are positively inviting danger, because as we have seen, their most-favoured organism *E. coli*, is something which lives in the human bowel. In other words, it is something which will breed and multiply in humans, and thus any deadly genetic intruders will breed and multiply likewise. As one critic asked: "To do potentially hazardous experiments, why pick on an organism that lives in us?" (Wald, 1976, p. 130.)

Undoubtedly, it was the prospect of recombinant DNA research killing us all that caused most of the furor. Were there no question of its safety, most of us would never have heard of it. But apart from the supposed dangers of recombinant DNA work, the critics challenge just about all the positive virtues claimed for such work: certainly all the unique virtues of such work. It is agreed, somewhat grudgingly, that recombinant DNA work is much faster and more powerful than alternative approaches, although there is some doubt about whether it is absolutely indispensible all the time: "Clearly there are experiments with recombinant DNA techniques that do not have readily apparent *in vitro* parallels. This does not mean that the questions such experiments hope to answer cannot be approached through research that does not require the manufacture of novel micro-organisms. The scientists wishing to conduct this research may simply have to be more patient" (SftP, 1978, p. 113).

However, as far as the supposed benefits are concerned, close examination shows them not to be so very beneficial at all. At least, the benefits can or will be purchased only at the expense of far more worthwhile ends. Consider the hypothesized medical virtues of recombinant DNA work. If one is really concerned to eliminate disease and improve health, in North America and

Europe not to mention the Third World, then one would do far better to concentrate one's efforts on preventative medicine and the like. Take cancer. At least 80% of cancers are environmentally caused, and half of these are due to smoking. How foolish to pour money into recombinant DNA research, finding a cure for cancer, when we know already what the cure is. Stop it in the first place! (SftP, 1978, p. 117)

According to the critics, similar sorts of arguments apply when we look at the supposed agricultural benefits of recombinant DNA. Whom are they supposed to benefit? As things stand already we have quite enough food to feed everyone: the problems faced by peoples from so-called 'underdeveloped' nations stem not from inadequate technology and supplies today, but from economic and social causes. There is sufficient food to go around, if only it were distributed properly and if we were to avoid gross follies like the feeding of grain to animals in North America, when known-alternative methods of farming could provide much higher protein crops for all. We really do not need fancy new ways of fixing nitrogen from the air. What we need is more just management and distribution of what we can already do (Reutlinger and Selowsky, 1976; Wade, 1976).

Identical arguments have been levelled against all the other promised benefits of recombinant DNA research. Critics combine doubt as to the real prospect of recombinant DNA research, wishes that we would start doing better or more justly what we can already do, and fear that side-effects might cause more pain and problems than pleasure. Beware of the gods bearing gifts. Thus, with respect to the proposal to create an organism capable of digesting crude oil, George Wald writes: "It is pointed out that this organism could be very useful for cleaning up oil spill. Very true; but how about oil that has not spilled? − oil still in the ground, or on the way, or stored? Can this organism be contained, kept from destroying oil we want to use? Or will we need to begin to pasteurize oil?" (Wald, 1976, p. 131)

Overall, the critics conclude that the most simple analysis shows that we should have little or nothing to do with recombinant DNA research. It is dangerous and we do not need it.

7.5. DO THE BENEFITS OUTWEIGH THE RISKS?

The time has come to compare and contrast the various claims. Remembering that I am leaving until later some of the more philosophical questions, let us try to compare benefits with risks. We can start by looking at the positive case that has been made for recombinant DNA research.

First, we have the supposed medical benefits: cheap insulin, cure for sickle-cell anaemia, and the like. It will come as no surprise to the reader, particularly the reader of my last essay, to find that I am perhaps not quite as enthralled by recombinant DNA research in this respect as are its advocates and exponents. Judged by either utilitarian or Kantian grounds, I agree with the critics that if we were really concerned with health care, there are probably far better ways of spending our time and money than on recombinant DNA research. Just as with genetic counselling, one cannot deny that some will probably benefit from such research; their lives will be made happier, and this is undoubtedly a good thing. On the other hand, when one thinks of the vast numbers who die each year from lung cancer, in almost all cases a direct function of smoking, and when one thinks of how little governments and other bodies (including medical bodies) do to prevent or disuade people from smoking, one can hardly take too seriously pleas that the elixir of life will be found only if we spend large sums of money on recombinant DNA research. Perhaps I might make a compromise suggestion. In Ontario, the province directs large sums of money to tobacco farmers; would we not get double benefits if these funds were diverted into medical uses of recombinant DNA? (In this context see Epstein, 1976 a, b.)

Please note that I am not trying to take an extremist position, arguing that no medical good will come of recombinant DNA work. What I am simply querying is whether overall quality of healthy life will be improved by sinking large sums into such work, rather than by diverting such monies elsewhere. I would add also that, taking into account some of the aims of genetic counsellors, an element of confusion seems to work beneath the surface with respect to aims. If we now have the power to eliminate sickle-cell anaemia through pre-natal diagnosis, is it worth investing large sums in a quest for the cure? Admittedly, I have tried to highlight difficulties in using genetic counselling in just such a case as sickle-cell anaemia: but it might be worth asking what the ideal course of action would be, before we go rushing off in all directions, aiming simultaneously for both elimination and cure.

Turning to agriculture, I am inclined to think that probably a stronger case can be made for recombinant DNA work in this respect. Much of what the critics say is undoubtedly true. We could do far better with what we already do have than we do at present, and this applies not only to the Third World but even to ourselves. The colossal waste, bad distribution, unequal and unfair use of resources, and like phenomena, are plain for even the most dim-sighted to see. On the other hand, although it is somewhat of a cliché to say, it is undoubtedly true that the world, including the industrial part of the world, faces a crisis, specifically a crisis in energy sources (R. C. Valentine, 1978). Already, in the past decade we have had strong intimations of this fact. It is true that, particularly in North America, there is gross squandering of every resource, our own and those of others, and it is true also that much of the energy-availability fluctuation in the 1970s has been less a function of real immediate scarcity and more of market manipulation and of economic and political upheavals (as in Iran); but only a fool would think that real scarcity of conventional sources of fuel is not something which will happen very soon. Moreover, before left-leaning thinkers smugly conclude that this will bring North America to its knees, it must be noted that already the evidence is that the fuel crisis hits most strongly at the underdeveloped countries of the world. Bangladesh suffers more than we do, absolutely and relatively.

Given these facts, it would seem to me foolish in the extreme not to explore fully all possible options and opportunities to reduce everyone's dependence on oil and the like. If indeed recombinant DNA research does offer such hope, for instance through the development of nitrogen-fixing plants, then it would be morally wrong — both from our own perspective and from that of future generations — not to explore such possibilities as fully as we can. Such explorations should, of course, be consonant with other priorities; no one suggests that all solar-energy research should be dropped to work on recombinant DNA research. Parenthetically one might add, without intending to detract from what has just been said, that not everyone is quite so sanguine about the possibilities of producing efficient nitrogen-fixing plants as are some of the most enthusiastic researchers. To manufacture nitrogen products from the air requires energy, and it might well prove that so much plant energy would have to be diverted from actual crop production that the overall returns would be diminishingly small (Grobstein, 1979, p. 64). This is a word of caution; it is not an argument to suspend recombinant DNA work in agriculture. The returns may be great; we shall have to see.

With respect to other possible benefits from recombinant DNA research I would argue in much the same way as I have just argued for agriculture. I think we need desperately to put our present house in order; but this is not to deny the even greater need to develop new technologies, especially given the fact that the life-style of the twentieth century — one directly dependent on unlimited fuel supplies — is coming to a rapid end, give or take a decade or two. Admittedly, it is perhaps a moot point as to how desperately the world needs an oil-gobbling bug. A few less ships sailing under the flags of convenience might do much to reduce oil-spills; although in defense of such a recombinant DNA project, if indeed one could make an oil-eating bacterium I would have thought that it would not be beyond human ingenuity to make it so that it had only limited powers. Perhaps it and its offspring could gradually poison themselves from body products, so that they would pose no general threat to the world oil supplies? But this is all a little bit in fantasy-land. My main point is that basic research with an eye to technological benefits is needed as never before. It may be considered bad taste to mention this fact − especially in a collection of philosophical essays − but even if we in North America do not pursue recombinant DNA research, the Japanese are bound to do so! Is biological technology to go the same way as the automobile?

Concluding this section, let me make one final general comment. There is today a popular species of bad argument which, following one perceptive writer, might be called the "fallacy of post-industrial romanticism": "[It] equates the *natural* or unmediated with the *good* − as in natural foods, organic gardening, earth shoes, homeopathy, 'totally honest' personal relationships and herbal everything" (Strouse, 1974, p. 8). There is much wrong with our culture today, and much that is wrong is due to obscene misuse of technology and to industrial society generally: additives to foods, gas-guzzling cars; very problematical nuclear energy generators; and so forth (Shrader-Frechette, 1980). On the other hand, let us not deny that, thanks to technology we are all a great deal better off today than we would have been two hundred years ago. To use Hobbe's memorable phrase, life then for most people was nasty, solitary, brutish, and short. We have health, leisure, happiness − *because of technology*. Things are obviously going to have to change. But to hope or to pretend that we are all going to go back to dancing around the May-pole on the village green, is a naive and dangerous illusion − as well as being a pretty boring prospect.

The good future does not mean no technology, or even less technology. It does mean different use of technology; but that is quite another matter. For this reason, although I am empathetic to many of the complaints of the critics of recombinant DNA research − with respect to prospective medical gains I have especially grave doubts − fundamentally I suspect that the critics and I fall on opposite sides.

7.6. THE DANGERS OF RECOMBINANT DNA RESEARCH

But surely, this discussion is all a little bit ostrich like! How can one even discuss or think of benefits when, as the Sword of Damocles, recombinant DNA research hangs poised, ready to destroy us all? In any cost-benefit comparison, costs, at least potential costs, quite out-swamp any benefits, real or apparent.

Now, let me say outright, that even without recent relaxing of standards, I think that anyone who wanted to argue that dangerous micro-organisms could never escape from establishments with high levels of containment would be very naive indeed. Inductive evidence shows that nothing is ever *that* secure or safe: as noted, recently foot-and-mouth disease got outside the strictest barriers that anyone has so far erected (Wade, 1978b). And this is not to mention disasters or near-disasters of a similar type like Three Mile Island, all of which increase one's scepticism about safety claims involving dangerous substances or operations. Moreover, although it is true that molecular biologists have made some efforts to improve safety factors of their work, even by their fellows such efforts have not been received with unbridled applause. For instance, Roy Curtis created a form of *E. coli* which could survive only under the most artificial of conditions (Curtis, 1979). But about this organism, 'χ 1776' another worker has written that it "is inconvenient to use, and requires constant monitoring to verify its many biological properties. Unless extreme care is exercised in its handling, it is very likely to be replaced in culture by a rapidly growing mutant, or by an extraneous contaminant" (Novick, 1978, p. 97).

Let us, therefore, concede the possibility of the escape of dangerous organisms. One might think that this is all there is to the matter: recombinant DNA research ought to be banned, absolutely. Indeed, if one reads the writings of some of the supporters of recombinant DNA research, let alone the critics, one would think that there could be no further argument. For example, the very same critic of Curtis's work quoted just above, a man who nevertheless thinks that recombinant DNA research should be allowed to procede, happily dreams up 'scenarios', quite possible we are assured, where newly created organisms could destroy the whole world.

Let us consider *Clostridium botulinum* (the producer of botulin toxin – the cause of botulism, and the most potent toxin known) . . .

As you may know, the *clostridia* are obligate anaerobes, i.e., are able to grow only in the absence of oxygen, and they will not grow in an acidic environment. Consequently, they can contaminate relatively few types of food and, in addition, cannot multiply in the human intestine or in living mammalian tissues. One could conceivably introduce genes permitting growth in the presence of oxygen or in an acidic environment into such an organism, so modifying profoundly its ecological potential. Possible consequences of the ability to grow aerobically would be the ability to infect humans and other animals with dreadful consequences and the ability to multiply in habitats exposed to oxygen. A consequence of the ability of *C. botulinum* grow in an acid environment would be the lethal contamination of foods (such as pickles, preserved fruits, etc.) that are at present entirely safe. (Novick, 1978, p. 90.)

One has visions of some mad fiend poised to destroy the youth of America through the stealthy introduction of poisoned pickles into Big Mac's. Who can ever look comfortably at a hamburger again?

But obviously anyone can spin scenarios, whether they be molecular biologists or not. The question we want answering is how seriously we should take them. I suppose it is always possible that turning on the lights of Toronto at night might attract alien beings with all the warmth and good-feeling of Darth Vader, but I for one am not going to lie awake at nights worrying about the possibility. Nor am I going to worry unduly about the possibility that one more year's use of hydro-carbons will so increase the carbon dioxide content of the atmosphere that, greenhouse like, the world will heat up and we shall all be flooded out by melting poles. In other words, what we must recognize is that there are crazy or unlikely scenarios as well as reasonable ones, and we must ask ourselves how plausible the recombinant DNA research scenarios really are. At least, rather than letting our imaginations be the sole arbiters of the recombinant DNA question, we must see if there are more solid foundations on which to decide matters of safety.

I suspect that the right way to assess the real danger of recombinant DNA research is to divide the work into two categories: that containing projects deliberately in search of or likely to produce dangerous organisms (in the light of past experience), and that which does not contain such projects. There is bound to be some overlap, intentional or otherwise, but one can undoubtedly think of some plausible dangerous recombinant DNA scenarios, I would think recombinant DNA experiments on the smallpox virus would fall into this category. Moreover, having drawn a line conservatively in favour of safety, I take it that there will be little doubt in anyone's mind that recombinant DNA work deliberately likely to produce dangerous organisms ought not to be allowed by all and sundry. How much dangerous research, if any at all, should be allowed is perhaps arguable and a decision will, of course, have to rest on the perceived degree of danger and the ends one is after. Certainly, even if one cannot guarantee water-tight barriers one can insist on the best possible barriers, as of course we do already when scientists have to deal with dangerous organisms like the (unmodified) smallpox virus. I myself would have thought indeed that a case might be made that certain kinds of work ought not to be allowed to all, at least until we know far more about recombinant DNA.

We must accept therefore that there could be dangerous recombinant DNA work. It is important not to fall into what I shall label the 'thin-end-of-the-wedge fallacy', namely assuming that because there is dangerous work, all such work is potentially dangerous and ought to be banned. If one follows this line of reasoning, then one ends up in the absurd position that all science ought to be banned, because identical arguments can be lodged against any branch of science. One can certainly think up lots of dangerous experiments in physics which one would not allow. Similarly in the social sciences. Suppose someone suggested teaching all the children in Toronto or New York that humans of a certain kind (for instance, left-handed people) are evil. One might learn all sorts of interesting things about indoctrination and prejudice – but I doubt one would think the results worth the drawbacks. However, because one has the possibility of such experiments in the social or other sciences, one does not forbid all social or other science experiments. One judges each one or each kind on its merits. Similarly, the possibility of dangerous recombinant DNA experiments does not, in itself, justify a ban on all such experiments.

What then about non-intentionally dangerous recombinant DNA research? Should it be allowed, or are the various possible scenarios of things going wrong sufficiently plausible to call for a total ban? Although one may think my position to be a classic case of a fool rushing in where angels fear to tread, or rather of a philosopher rushing in where molecular biologists fear to tread, I rather suspect that the dangers of normal recombinant DNA research have been overdrawn and that, in fact, there are no justifiable grounds for a ban. I hasten to add that I argue as I do, not because I think I am wiser than molecular biological critics of recombinant DNA research, but because I doubt that they are quite as wise as they think they are (or rather, thought they were). The simple truth is that because someone is a molecular biologist, it does not follow that he/she knows very much at all about the spread of disease! It may be fun and attention-catching to spin scary scenarios, but the ability to cut up a gene is not equivalent to professional ability to judge potential dangers from micro-organisms. For this we need to turn to professionals in the area of disease and its spread, that is to epidemiologists. As one expert in this field has written:

Instead of recognizing that the familiar features of the organisms offered grounds for reliable, general epidemiological predictions even before laboratory studies on the new strains provided more precise information, the scientists involved proceeded as though nothing relevant was known and they were peering into a black box. The resulting preoccupation with unlimited scenarios contributed to the anxiety that they came to feel − and then transmitted to the public. To be sure, in order to establish the patho-genicity of any novel recombinants with precision, direct tests would be necessary. But a purely empirical approach could not be sufficient, for the next experiment might still inadvertently unleash an Andromeda strain. Eventually, then, it would be necessary to take into account also what is know about epidemics − and this information could have been helpful from the start. (Davis, 1977, p. 549.)

This note of common sense, incidentally, is made all the more pertinent by the fact that the writer has been cited as an expert in the field by the very critics of recombinant DNA research! (Chargaff (1976) referred to Davis's text-book on microbiology in support of his claim about the folly of using *E. coli* as the chief organism in recombinant DNA work.)

Let us, therefore, follow our own prescription and turn to epidemiology, seeking help with respect to our queries about the dangers of recombinant DNA research: any non-intentionally dangerous recombinant DNA research.

7.7. THE ARGUMENT FROM EPIDEMIOLOGY

Apparently, we can locate three points of potential risk: first, that recombi-nant DNA research will produce a potentially dangerous organism; second, that such an organism will infect a laboratory worker; and third, that such an organism will spready beyond the laboratory. We consider these three points briefly in turn.

As far as the first point is concerned, the question is whether inadvertently dangerous organisms might be produced, for instance through 'shotgun' experiments. In the light of earlier discussion, the possibility of human tumor virus genes being incorporated in *E. coli* would seem to be particularly worri-some. *E. coli* live in the human bowel, and the last thing we would seem to want are organisms which can live in humans and which contain chunks of cancer genes; although we hardly want any other kind of dangerous gene either.

However, for a number of reasons the fears raised at this point may not be that justified (Davis, 1977, 1979a; Formal, 1978; Freter, 1979). Probably tumor viruses are going to be rare − if so, the chances of being incorporated are slight. If they are common, then presumably a few more will not make that much difference! Moreover, there is probably no such thing as a cancer virus *per se* − the problem is that under certain circumstances certain genes give rise to cancers − these genes in other circumstances are 'normal'. In other words, even if we got a certain gene into an *E. coli* and then into a human, there is no reason to think that it would in itself be particularly dangerous, either in the *E. coli* or transferred back to a virus. And finally and most importantly, a point which counts against 'cancer' genes or any other kind of dangerous gene, shotgun 'experiments' analogous to human research, are probably going on in nature all the time, anyway. DNA in the gut, released from dead cells, will sometimes be picked up by the bacteria there. Hence, in the human case, there will have been selection to protect us against bacteria containing chunks of human DNA. In short, with respect to the first point, the problem of dangerous organisms, research within safeguards seems fairly safe.

Moreover, it seems now that there is, as it were, a double safeguard against the dangers of shotgun experiments, or any other kind of experiment which might introduce a dangerous piece of human DNA into a bacterium, which latter might then multiply and cause further damage all around (Wade, 1979a). Not only are we humans selected for natural protection against bacteria containing new genes, but there is reason to think that genes from higher organisms incorporated into bacteria simply do not work anyway! At least, lest it be feared that this point backfire on all the technological virtues that are hoped for from recombinant DNA research, it is thought that with suitable manipulation the genes can be made to work, but that special non-natural conditions are needed. In other words, there is no question that the products of shotgun and like experiments will at once be rearing to go out and do untold damage on us humans, even if they could. Nature has a second barrier of protection for us.

We come to the second point of potential danger in recombinant DNA research: possible illness in a laboratory worker. As we know, it is at this point that the opponents of recombinant DNA research start to get really frenzied, because they feel that the choice of *E. coli* on which to do experiments is playing with fire. "If our time feels called upon to create new forms of living cells — forms that the world has presumably not seen since its onset — why choose a microbe that has cohabited, more or less happily, with us for a very long time indeed?" (Chargaff, 1976, p. 938). However, for at least two reasons, this apparently reasonable objection does not have the force one might fear.

First, in one very important sense *E. coli* is a very safe organism to work with. Instead of floating through the air and then infecting people — something which requires very careful precautions against infection — it has to be swallowed. Swallowing is much less likely to occur accidentally. Second, the *E. coli* of the laboratory today, *E. coli* K 12, is just not the *E. coli* of the human gut. In its fifty years in the laboratory, it has been selected right away from its ancestors. And what is most important, the selection which has occurred making it ideal for laboratory use, has been at the expense of adaptation to the human bowel. *E. coli* K 12 simply does not last in humans. Within a day or two of swallowing huge doses, far more than one might expect in a laboratory accident, all traces vanish from human stools. In other words, it is quite misleading to object that most recombinant DNA work goes on in an organism indigenous to humans. "*E. coli* K 12, outside the laboratory, is like a hothouse plant throw out to complete with weeds" (Davis, 1977, p. 552; also compare Tables 7.1 and 7.2).

TABLE 7.1

Duration of shedding ingestion of *E. coli* strain hybrid strain by healthy volunteers

Volunteer	Dose (cells)	Duration of excretion (days)
1	1×10^8	12
2		0
3		16
4		60
5		45
6	1×10^{10}	105
7		21
8		7
9		14
10		75

(Formal, 1978, p. 128)

TABLE 7.2

Duration of shedding following ingestion of *E. coli* K12 – Shigella flexneri hybrid strain

Volunteer	Dose (cells)	Duration of excretion (days)
1	1×10^8	0
2		0
3		3
4		4
5		3
6	1×10^{10}	0
7		5
8		4
9		4

(Formal, 1978, p. 130)

Finally, we come to the third and perhaps most important point of all. What are the risks of escape from the laboratory and of an epidemic? Again, in the opinion of reading epidemiologists, we should not overestimate the dangers and raise fears. By adding more DNA to a *E. coli*, we are not going to turn it into something other than it is, namely an *E. coli*! (We are augmenting the DNA by about 0.1%.) Hence, if the organism is to spread and cause disease, it has to spread and cause disease in the way that *E. coli* spreads and causes diseases, namely through the water supply and the like (i.e., it will not suddenly become airborn like influenza). But, already we have good sanitation techniques designed to stop the spread of *E. coli* and similar organisms — chlorine in the water for instance. In other words, the chance of danger from a virulent form of *E. coli*, even if it escaped, is slight indeed.

So what conclusions can be drawn? Simply that, taking all things into consideration, the overall dangers from normal recombinant DNA work appear "to be vanishing small, compared with the very real danger faced continually by medical microbiologists working with known virulent pathogens — organisms well adapted to survive and produce disease" (Davis, 1977, p. 553). What more can be said? On the one side we have, or at least had, all kinds of frightening pictures drawn up, limited (and justified) in their scariness only by the human imagination. On the other side we have (another cliché I am afraid) the voice of experience. The only reasonable conclusion seems to be that the fears of recombinant DNA research and its consequences were not justified. In the light of what we know, normal recombinant DNA work seems to be essentially without danger. And this being so, we can conclude our cost-benefit discussion of recombinant DNA research by agreeing with its supporters that on balance prospective benefits do seem to outweigh actual and prospective costs. How valuable it will prove to be technologically remains to be seen, but it does certainly seem to be an avenue well worth exploring.

7.8. RECOMBINANT DNA RESEARCH CONSIDERED AS SCIENCE

Deliberately so far I have kept the discussion at the pragmatic level of costs and benefits: what will recombinant DNA research do for us? What might such research do to us? But this is to tell only part of the story. Humans do not live by bread alone. The glory of being a human being, and it is a glorious thing to be, is that we rise above our brute existence. There is more to human life than simply scratching out a living. We have the urge to discover and to create. We want to know why things are as they are, and we want in lesser or greater ways to add to human existence — whether it be merely through a pretty flower-bed or through an overwhelming work of art.

What I would argue is that science — pure science for its own sake — is no less part of this most noble of human urges and achievements than is any one of the fine arts or of the other enterprises which drive humans beyond the level of the brutes. As a testimony to human greatness — proof that although we may be little higher than the apes we are nevertheless little lower than the angels — the theory of the double helix ranks right along with *Don Giovanni*, the funeral mask of Tutankhamun, the Taj Mahal, and Plato's theory of forms. When all the practical considerations are stripped away, and do not think that I am belittling these, science is a beautiful thing in its own right, and it is to be cherished for this very reason. Thus, I would argue that recombinant DNA research, which seems to be science at its most exciting and most fruitful, is a good and ought to be encouraged and praised. It has intrinsic worth which justifies itself.

I hope I will not be misunderstood. I am not arguing that the recombinant DNA researchers ought to be given complete license or that they have no social responsibilities. Of course they have. Such work ought not to be pursued where general harm might result, and the taxpayers who support so much of the scientific enterprise today obviously have the right to insist that basic research not get a disproportionate share of limited funds. What I am pleading is that we not forget the value of science itself. And this I fear it is easy to do, even though one often sees brief lip-service paid to that ideal. Unfortunately, much of the neglect stems from the nature of science itself, particularly as it is today (Rescher, 1978). One gets far larger grants from government and other agencies if one can suggest technological benefits from research; so these ends tend to get stressed. Also science today is a cooperative endeavour, done in a professional manner — gone is the supposed romantic individualist like Charles Darwin working away in his home in Down. Hence, the sheer beauty of the creativity and the discovery gets masked. But the beauty and the joy it inspires is there nevertheless, together with the ambitions and other motives which drive and inspire scientists; so perhaps I, as a philosopher, may be allowed to highlight and extol it.

Some I fear will object that I am too much of an idealist. Let me look at three objections to my stance, specifically as it applies to recombinant DNA research. The first objection is one based on a belief which in one form or another has been around at least since Herbert Spencer (i.e., 1850s), and which keeps cropping up in the writings of evolutionary biologists, most recently in the work of the sociobiologists (more on them and their ideas in the next essay). The belief is that, in some sense, the course of evolution dictates or shows what is morally good or worthy, and the consequent objection is that it would be wrong of us to try to change the natural course of evolution, specifically through the creation of new organisms or forms by means of recombinant DNA techniques. Hence, any moral or aesthetic worth which comes from recombinant DNA research considered as pure science is outweighed by the moral transgression of changing the natural course of evolution.

Arguing somewhat in this vein we find one critic of recombinant DNA research, Robert Sinsheimer, saying that:

There is also, one suspects, a profound innate wisdom in the diversity of the extant human gene pool, evolved and refined by trial and error in the crucible of natural selection. Myriads of our ancestors suffered bitterly to bequeath us this treasure. Diversity is the source of creative interaction and the seed of future adaptation. We should not lightly replace such an intricate pattern with the designs of human ingenuity. (Sinsheimer, 1978, p. 26)

Admittedly Sinsheimer does concede that some improvements in nature may be possible: but essentially his is a conservative position, looking upon the actual as the good. "Once launched upon genetic engineering we are forever on our own. Dare we introduce such an historic change?" (ibid, p. 27) Similarly, Chargaff asks rhetorically: "Have we the right to counteract, irreversibly, the evolutionary wisdom of millions of years, in order to satisfy the ambitions and the curiosity of a few scientists?" (Chargaff, 1976, p. 940).

The answer to Sinsheimer and Chargaff and like thinkers is that there is nothing in the course of evolution *per se* which tells us about right and wrong, about what we ought or ought not do (Ruse, 1979b). It is absolute nonsense to talk about the 'evolutionary wisdom of millions of years' — or at least it is pertinent to point out that this evolutionary 'wisdom' gave us smallpox, tuberculosis, the plague, and tobacco. It is certainly true that as far as humans are concerned, generally it gives us pleasure to pay attention to our bodily needs and functions like food, sex, and warmth. Moreover, satisfaction of these needs and functions has evolutionary significance, and no doubt the pleasure we get from the satisfaction is, by and large, a product of evolutionary processes. And we accept that pleasure and happiness are things worth striving for, things with moral worth or value. But we do not derive the values from our bodily functions; rather we evaluate the functions and our satisfaction of them in the light of values. From an evolutionary perspective it might be biologically adaptive to be terribly promiscuous, but it does not follow that such promiscuity is a morally good thing. Far from it.

Analogously, one can certainly say that one has the right, if not the obliga-
tion, to evaluate recombinant DNA research in the light of one's fundamental
moral principles. Will it bring happiness? Will it respect individual rights? (see
Essay 6). But these are the only kinds of moral questions one can and should
ask. There is nothing else. Evolution is just a fact. If one argues otherwise,
that evolution is in itself a good thing, then one has to conclude that the
efforts of the WHO in eliminating smallpox were immoral, because it was
making a certain species go extinct. And this is ridiculous. One is committing
what philosophers term the "naturalistic fallacy", namely one is trying to
derive moral claims from factual claims (Moore, 1903; Flew, 1967). Thus,
the sweeping questions and statements of Sinsheimer and Chargaff pose no
genuine threat to recombinant DNA research.

A second objection is one to be found in the critique of recombinant DNA
research by a group of radical scientists from Boston, the 'Science for the
People' group. As part of a general attack on any and all attempts to provide
biological foundations for human behaviour — an attack we shall consider in
some detail in the next essay — they object that recombinant DNA research
can and will encourage efforts to give genetic explanations for intentions
and actions.

In advancing recombinant DNA technologies as means to understand and deal with a
great variety of problems, the proponents of these technologies are contributing further
to this emphasis on genetics in social policy. This will have the consequence that programs for providing services that change the enviornment will be neglected in lieu of
genetic assessment; attempts at understanding the interaction of environmental and
genetic factors in dealing with social problems with the intent of best serving people as
they are will be deferred in favor of developing systems that track or channel individuals
based on their genetic limitations. (SftP, 1978, p. 119.)

I empathize strongly with many of the objections and queries that the SftP
group have about recombinant DNA research. Nevertheless, I would argue
that at this point they overstep the mark. Leaving on one side the whole issue
of putative biological foundations for human behaviour and the consequent
relevance of biology for decisions about social policy — these will be the
direct subjects of discussion in the next two essays — their objection to
recombinant DNA research seems unfair, considered both from a limited and
from a general perspective. The particular unfairness is that (unlike sociobiology, the SftP group's *bête noire*) there is no emphasis in recombinant DNA
research on behaviour as opposed to other organic attributes, human or
non-human. Certainly recombinant DNA work could impinge on human
behaviour, as for instance, if one tried to cure some genetic disease affecting

behaviour (like certain forms of schizophrenia) by using therapy based on recombinant DNA techniques. But most of the work, actual and proposed, has little to do with human behaviour. Indeed, much has little to do with humans directly, at all. Hence, there is no reason to think that recombinant DNA research specifically is going to lead to a biological view of human actions.

But this leads to the general unfairness of the SftP's criticism. Suppose one gets the likely response that the real danger is that recombinant DNA research contributes to an overall tendency to think the answer to human dilemmas lies in biology, and that this is dangerously wrong. Granting what I have already granted, namely that one ought not think that technology *per se* offers a simple solution to all human worries, this objection is surely unfair because, if taken seriously, one would have to proscribe all work in biology! The conventional animal and plant breeder, the researcher on Drosophila, the normal molecular biologist, are all just as guilty of leading one to think biology important and to increase (or at least continue) reliance on biological technology. Suppose, however, the critics agreed with my argument, but took the correct conclusion to be, not that recombinant DNA research should be spared, but that perhaps all biological work should be stopped, or at least heavily monitored for ideological priority. In a sense I feel somewhat helpless in the fact of such an absolutist stance. However, it is surely worth pointing out that, apart from the hypocrisy that such a stance would force upon the SftP group because many of them have themselves contributed significantly to the progress of biology (Richard Lewontin is a member!), a similar argument would seem to hold against all other sciences. Certainly all of physics stands condemned because we all know of the dangers of nuclear technology, and the same holds for the social sciences because they too can be used for evil ends (as in brain washing). In short, if one is not to pick unfairly on biology, one must proscribe all sciences. And surely such a conclusion really does get very close to being a *reductio ad absurdem* of the initial argument. Biological technology has its dangers. But the solution lies not in intellectual ludditism, refusing to have any part in technology or in a science which might be turned to technological ends; rather we must show a more subtle appreciation of the virtues and limitations of science and technology, present and future.

7.9. CAN ONE REALLY SEPARATE SCIENCE AND TECHNOLOGY?

The third objection to the case I have tried to make, defending the value of recombinant DNA research simply on the grounds that it is pure science and thus a good in itself, is one which has been made by the distinguished philosophical commentator on the contemporary scientific enterprise, Hans Jonas (1976, 1978). Essentially what Jonas does is to challenge the distinction I have drawn between pure science and technology, thus (if his challenge is well-taken) entirely undercutting my whole line of argument. Although Jonas does not think that the distinction is a logical absurdity, perhaps indeed it could actually be made, back in Greek times, he does feel that the distribution no longer is viable. The rise of modern science has "entirely altered the traditional relation of theory and practice, making them merge ever more intimately" (Jonas, 1976, p. 15). Indeed, "not only have the boundaries between theory and practice become blurred, but the two are now fused in the very heart of science itself, so that the ancient alibi of pure theory and with it the moral immunity it provided no longer holds" (Ibid, p. 16). Particularly since the Industrial Revolution, scientific advances have been spurred by and evaluated in terms of their potential technological payoffs.

Therefore, to Jonas the science-technology distinction is no longer a really valid one. All science now is essentially a kind of glorified technology. Moreover, argues Jonas, we have further evidence that science has come to be a kind of technology because of the nature of its aims: " . . . it has come to be that the tasks of science are increasingly defined by extraneous interests rather than its own internal logic or the free curiosity of the investigator" (ibid, p. 16). And we should note that Jonas is quite explicit in his belief that recombinant DNA research violates or merges the sharp divide between science and technology. "Clearly . . . recombinant DNA research on microorganisms alone does fit with a vengeance our . . . description of how science has forsaken the sanctuary of pure thought and become enmeshed with action that is subject to extra-scientific criteria of the public good" (Jonas, 1978, p. 265). Hence, no defense of recombinant DNA as a good in itself is possible. Such a defense is based on a false premise.

Of course, even if one grants Jonas's argument, which I shall not, one would still be able to defend recombinant DNA as research if benefits outweighed accepted costs. I suspect in fact that Jonas and I stand on opposite sides as far as this question is concerned; but I have made my case in this respect and shall let it rest without further comments. The important question for us here is whether in pointing out that science has practical applications, Jonas has negated my major point that one can still talk meaningfully of pure science for its own sake, and that, specifically, one can cherish recombinant DNA work because it is an instance of this. I do not see that Jonas's argument touches my claim. Whatever the putative practical implications of recombinant DNA research — and I have admitted these fully — it is still a genuine scientific activity producing genuine science. It is not merely a sophisticated technological endeavour, and therefore only worthy to be evaluated on those grounds.

Jonas's only substantial argument for this position rests on his claim that "the tasks of science are increasingly defined by extraneous interests rather than its own internal logic or the free curiosity of the investigator" (Jonas, 1976, p. 16). But this is just not necessarily so, even today; and it is particularly not so in the case of recombinant DNA research. To be perfectly honest, I am not quite sure what the 'internal logic' of science really is, but if most obviously we look at history of science for guidance, then what we see is an increasing effort to bring the world beneath natural law. Moreover, although vitalists do not much like this fact, we see effort to connect everything up, particularly in the sense that larger things are explained in terms of smaller things, which usually involves older theories dealing with things at a gross level (like Boyle's law) being shown to be consequences of newer theories dealing with things at a more detailed level (like the kinetic theory of gases). That is to say, one sees a kind of *reductive* trend in science. In physics, macrophenomena are explained in terms of microphenomena. Chemistry is explained in terms of physics. Macrobiological phenomena like organic characteristics are explained in terms of smaller things like genes. And today we have a major controversy as the social sciences are being brought kicking and screaming into contact with the biological sciences, the controversy over so-called 'sociobiology'. (See the next two essays, as well as Essay 5.)

Now, whatever one might think of it, recombinant DNA research stands about as firmly in this tradition as it is possible for something to be. What we have is inquiry on the borderline between the physical sciences and the biological sciences, as scientists are finding out in the most basic way, precisely what makes the world work, seeing if in some manner the gap can be bridged between the inorganic and the organic domains. Can organic macrophenomena, like phenotypes, be explained in terms of inorganic microphenomena, like molecules? If this is not a case of science driven by its own internal logic, I do not know what is! Moreover, since the whole area of inquiry stands at such a key point on the overall scientific reductive plan, it would be ridiculous to deny the genuineness of the inquiry of the scientists concerned. This is inquiry in the most fundamental and mainstream sense. Only the most bigoted could say that there is nothing but technology and putative practical applications at issue here.[2]

In short, what I argue is that recombinant DNA research does count as scientific activity in a proper sense and is therefore, in its own right, worth cherishing. It is science absolutely at the cutting edge. Hence, Jonas's argument fails, as do the other theoretical objections to recombinant DNA work.

7.10. EPILOGUE

So what can we say, looking back and looking forward, now that the recombinant DNA furor has subsided? Admitting total defeat in my struggle to avoid clichés, I would suggest that there certainly is no cause for complacency but that things could be a lot worse than they are! Undoubtedly silly and ill-thought-out things were said by people on both sides. Equally neither side was free of makers of exaggerated claims. However, concern was shown by the very people most intimately involved in recombinant research, and matters were debated freely and publicly until reasonable and informed estimates of costs and benefits of such research started to emerge. That, it seems to me, earns the scientific community a very large moral plus. It would be foolish to think that there is no further need for caution, but as we start into the second decade of recombinant DNA work, justifiably the main emotion can be one of excitement not fear.[3]

NOTES

[1] I hope no one will take me as implying that this means that molecular biologists' achievements are really quite ordinary. God forbid! I stand firmly with Whewell in believing that simplicity is an essential — perhaps the overriding — mark of the greatest of all theories. To find simplicity in this complex world of ours is a wonderful achievement.

[2] Obviously here I am presupposing the kind of logic of reduction which I defended in Essay 5. However, even if one thinks that what I have characterized as 'replacement' is more common in science than I allow, the 'internal logic' of science still seems to involve physico-chemical notions moving in to explain biological phenomena. Hence, recombinant DNA research still seems to be central to the whole enterprise.

[3] The whole 19 September 1980, issue of *Science* (209, 4463) is devoted to the present state of recombinant DNA work. It provides an excellent, although necessarily technical, survey.

SOCIOBIOLOGY: SOUND SCIENCE OR
MUDDLED METAPHYSICS?

The public image of science is that of an enterprise of dispassionate, cool objectivity: something involving sober, emotion-free, white-coated people, as they battle collectively to wrest secrets from the stubborn universe. Nor is this an image entirely unacceptable to scientists themselves. How else can one explain the great fondness they have for Sir Karl Popper's philosophy of science, with its picture of men of science ruthlessly discarding favoured brain-children in the face of ugly but falsifying facts?[1] As we know, even biologists avidly swallow and regurgitate this myth, despite the fact that Popper calmly tells them that their most important theory is but a collection of half-baked truisms (see Essay 3). However, as historians of science know only too well, much of the actual activity of science descends right down to (metaphorical) bare-knuckle fighting of the most bloody kind (Ruse, 1979a). Today, there exists just such an acrimonious scientific dispute over the supposed new discipline of 'sociobiology', an area of inquiry concerned with biological causes of animal social behaviour, including human social behaviour. On the one hand, some absolutely first-class biologists feel that it represents a basic breakthrough in a crucial area of evolutionary thought. On the other hand, some equally first-class biologists feel that in important respects it represents an excrescence on evolutionary science: an excrescence of the most reactionary and malignant nature.

It might seem improvident, not to say impertinent, for a non-biologist to think of plunging into the fray and of commenting on, what from politeness we might call, the sociobiology debate. Nevertheless, foolhardy though such a move might be, it is not entirely inappropriate for a philosopher to consider the controversy. As so often happens when scientists fall out, much of the disagreement centres not on matters of brute scientific fact (whatever that might be), but on questions essentially philosophical: metaphysical and methodological (Kuhn, 1970). This being so, I intend to present in this paper an almost autobiographical account of how one, with philosophical training and a deep interest in biology, approached and reacted to the sociobiology controversy. I shall assume virtually no prior knowledge about sociobiology, being armed with little more than a very great respect for the biological achievements of some of the leading protagonists in the debate. I am really

190

sure, therefore, of only one thing: namely, that whether I am to end by agreeing or disagreeing with various of the positions taken, by virtue of prior work the people holding these positions have earned the right to have their opinions treated with care and sympathy.

To accomplish my end I begin with a sketch of what I take to be the major tenets of sociobiology, paying particular emphasis to those claims which have caused most controversy. Then I turn to the critics' objections, particularly those objections to human sociobiology, trying to assess their forces. And I conclude with a few brief reflections of my own about certain matters where, supposedly, sociobiology can provide answers to hitherto insoluble philosophical questions. (The material on sociobiology is already massive and growing exponentially. Rather different but good introductions are Dawkins (1976) and Barash (1977). A spirited criticism is Sahlins (1976). Good collections containing arguments both *pro* and *con* are Caplan (1978), and Barlow and Silverberg (1980). A brilliant essay in human sociobiology has been produced by that most entertaining of writers, van den Berghe (1979). Finally, I shall not let modesty deter me from mentioning a work which has found rather more favour among the sociobiologists than among their critics, Ruse (1979b).)

8.1. WHAT IS SOCIOBIOLOGY?

Clearly the place to start is with E. O. Wilson's major work, *Sociobiology: The New Synthesis* (1975a). If only by virtue of its massive size, it deserves to be called the 'Bible' of the new subject. Because it is this work in particular that has attracted so much hostile attention and, in order to keep this essay's size within bounds, the reader is warned that I shall concentrate on *Sociobiology* almost to the entire exclusion of direct consideration of work by others. This act is not quite as henious as it might seem, for as the full title of Wilson's book rather hints, Wilson is as much a collector and synthesizer of the ideas of others as an innovator in his own right. Recently, Wilson has returned to the fray with another work on sociobiology: *On Human Nature* (1978). In my next essay I shall be looking specifically at some of the claims in this volume, although I think one can say fairly that much that is expounded in detail in the later work is presented in capsule form in the earlier work. Even in this present essay it will prove convenient to make occasional reference to the later work.

'Sociobiology' we learn "is defined as the systematic study of the biological basis of all social behaviour. For the present it focuses on animal societies, their population structure, castes, and communication, together with all of the physiology underlying the social adaptions. But the discipline is also concerned with the social behaviour of early man and the adaptive features of organization in the more primitive contemporary human societies" (Wilson, 1975a, p. 4). So far, so good. What we seem to have, therefore, is a kind of general sociology-cum-anthropology. Not just a sociology of humans, but also one descending down (or, given one's viewpoint, ascending up) to primates, birds, fish, insects, and what have you. Where therefore lies the originality (and consequent controversial aspects) of sociobiology?

The answer is not hard to find. Sociology (of humans) is a discipline carried on, by and large, at what biologists call the 'phenotypic' level, the macroscopic level of the whole organism, its features and its behaviours. Sociobiology, however, is going to look also at the other side of the biological coin − the genotypic level. (See Essay 1 for more on the phenotype/genotype distriction.) In particular, sociobiology is going to look at animals and their behaviours, at least in part, from the viewpoint of their genetic makeups; and since the whole business of genes invariably raises questions to do with evolutionary origins, more generally one can say that sociobiology will attempt to understand animal behaviour (including human behaviour) in the light of our knowledge of evolutionary processes. As Wilson says truly: "This book makes an attempt to codify sociobiology into a branch of evolutionary biology and particularly of modern population biology" (Wilson, 1975a, p. 4). Since evolutionary theory is apparently so important to the sociobiological enterprise, it will be worth our while to pause for a moment and to recollect the modern 'synthetic' theory of evolution, that combination of the selectionist ideas of Charles Darwin and the ideas about particulate modes of inheritance of Gregor Mendel and his successors. Having done this, we can then more easily discern the theoretical structure of sociobiology, and its relationship to other parts of biology. (For more details of the synthetic theory refer back to Essays 1 and 2.)

Remember that the modern evolutionary theory starts with the idea of the gene, something on the chromosomes, which can be passed on from generation to generation, and which is responsible (with the environment) in some ultimate sense for the gross phenotypic characteristics, both static and dynamic. By virtue of neo-Mendelian principles one can show that, all other things being equal, genes in a largish population have a stable balance − with calculable rations encaptured in the so-called 'Hardy−Weinberg' law. Of course, 'all other things' never are equal, and this is where 'population geneticists' really set to work. Several 'forces' are thought to affect gene-ratios, and these thence lead to change, or, in certain circumstances, to different kinds of balance. There is immigration and emigration in and out of populations; there is mutation from one gene form to another; and most importantly, there is the fact that genes cause their possessors to reproduce at different rates, this differential reproduction being the Darwinian component of 'natural selection' (Dobzhansky *et al.*, 1977; Ayala and Valentine, 1979).

Population genetics thus deals with gene ratios in groups of organisms and with the causes of change of these ratios, and it seems clear that this, writ more or less large, is what evolution is all about. Considered over periods of time, particularly large periods of time, we get genetic changes, and this is organic evolution. Population genetics is, therefore in an important sense, the 'core' of evolutionary theory, and what I have suggested in earlier essays is that it binds together and provides the theoretical underpinnings of all the various disciplines which loosely come together under the heading of 'evolutionary studies' — paleontology, biogeography, evolutionary systematics, embryology in a sense, and so on. All of these disciplines, or subdisciplines, have their own peculiar assumptions and borrow freely from each other — but it is population genetics which gives them their theoretical backbone and unity. (See Figure 1.3.)

Now, returning to sociobiology, it seems that what Wilson wants to do is to add another sub-discipline to the evolutionary picture as I have portrayed it (Figure 8.1). Indeed, if one wanted a paradigm to illustrate evolutionary theory as I have characterized it, Wilsonian sociobiology would be it — in intention at least. Early in his work we get a detailed discussion of population genetics, beginning with the Hardy—Weinberg law, and then going on to more advanced and sophisticated results. Then these theoretical ideas are used to underpin and illuminate the discussions about animal nature and behaviour in

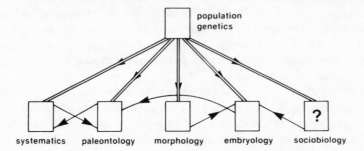

Fig. 8.1. How Wilson sees the proper relationship between sociobiology and the rest of evolutionary theory. (As in Figure 1.3, the lower level links are just illustratory.)

the following pages. I have argued earlier (in Essay 1) that we can properly regard evolutionary theory as hypothetico-deductive, in sketchy outline at least. Suffice it to say here only that sociobiology will probably not shift at all the impass between me and those who reject my analysis. The early parts of the theory, the populational genetical parts, are more or less deductive. From then on though, as so often happens in evolutionary work, the inferences get pretty loose, particularly when it comes to questions of how particular genes might aid their possessors in the struggle to survive and reproduce — that is to say, when it comes to questions to do with the relative 'fitness' of organisms due to their genes and with the 'adaptations' possessed by organisms at the phenotypic level because of these genes.

Nevertheless, in fairness to Wilson (and those upon whom he draws), it is important to emphasize that what he may lack in tightness of argument is more than compensated for in breadth, interest and sheer daring, as he attempts to make the study of social behaviour a fully fledged member of the evolutionary family. We learn how it is that certain species of African insect are able to control the heat and humidity of their nests, with a precision that out strips the ability of any modern engineer, and we discuss the adaptive value of such ability. We look at the behaviour of birds and why it is that, unlike mammals, male birds so frequently get involved in child care. We consider the life of lions at work and play: mainly work for the females, mainly play for the males. We find out why it is that certain species of animals, for example dogs, customarily hunt in packs rather than singly, and we look at the implications, benefits, and liabilities of being a group member rather than a loner. We examine the childhood, adolescence, and adulthood of primates — when, if ever, would a dominant adult male be prepared to allow another male near or into his balliwick? And thanks to the light that Wilson throws on the living, breathing, fighting, playing, eating, defecating, courting, copulating, parenting, dying animal world, we learn of and explain many, many more things, from the common place via the esoteric to the frankly absurd. Did you know that bands of wild male turkeys wrestle to establish top turkey? Or that female hyenas have pseudo-penises so that they too can join in 'show and tell'?

Some of the facts may sound somewhat incredible, and one may not always agree with Wilson's attempts to relate certain behaviours to adaptive advantage, but as far as the general enterprise is concerned, namely using evolutionary theory to explain animal social behaviour, it all surely sounds fairly uncontroversial. Moreover, many of the sociobiological explanations have at least an initial plausibility. Thus, take the fact that male birds get involved in offspring-care. The explanation given for this is that birds tend to rear their young in trees whereas mammals rear their young on or under ground. Life is more precarious up a tree than in on/under ground, particularly during winter. Hence, there is strong selective pressure on birds to grow and mature rapidly, and an obvious aid in this is the presence of two caring parents rather than one. Consequently, father gets involved.

In short, looking at the bulk of Wilson's argumentation, unless one sees red at the very mention of evolution, there seems nothing very much to get upset about. After all, analogously, to the best of my knowledge, no one was desperately perturbed when in *Tempo and Mode in Evolution* (1944), G. G. Simpson did for paleontology what Wilson is attempting to do for animal social behaviour. Where the trouble starts, of course, is in Wilson's attempts to apply his ideas to our own species. And although only some 30 pages (out of 700) are devoted directly to humans, and although the book is in no way a polemic with everything else merely preparing the way for humans, Wilson does rather lead up to the discussion of humans as a kind of climax to the book. Certainly it is this that the critics have fastened on, and so it is to this that we must turn almost at once in order to conclude our direct exposition of Wilson's ideas. But before so turning, better to flesh out our discussion of Wilson's general position (and with an eye to the future) it is important to expand a little on a topic in the animal world that Wilson treats at some length.

In his discussion of the principles of population genetics Wilson spends much time in expanding on the ways in which natural selection operates. When one thinks of natural selection one thinks normally of individual selection − in this sense it is an entirely selfish phenomenon, in that selection works to preserve characteristics helpful only to the individual. Suppose that in a particular generation one has two organisms, A and B, and that these have genes a and b giving rise to characteristics α and β. Say also that, by virtue of having characteristic α rather than β, A beats out B in some way (skills, drives off, matures more rapidly), and thus A breeds whereas B does not. We have a differential reproduction, and in the next generation we have just a genes. We say that a genes have been 'selected' over b genes. (See Essay 3 for more details on natural selection.) The trouble with natural selection construed only in the way is that it provides absolutely no explanation of *altruism*, a phenomenon which Wilson calls "the central theoretical problem of sociobiology" (Wilson, 1975a, p. 3). Why is it that we find so often in the animal world that organisms help other organisms (even of different species)? Individual selection implies that every organism is for itself. Moreover, it would seem that some kind of 'group selection' mechanism — supposing that selection can work simply for the benefit of the group — would not explain altruism, if only because group selection probably could never work. Suppose one had two organisms, one with characteristics of value to itself and the other with characteristics of value to the group. The selfish organism would easily out-reproduce the altruist, because both would be working for it and not for the altruist! (Williams, 1966).

The answer proposed to resolve the paradox of altruism — an answer which is certainly not totally original with Wilson or other modern-day sociobiologists, for in a sense, it goes back to Darwin (Ruse, 1980d) — is that altruism is basically enlightened self-interest (West Eberhard, 1975). In helping others, one helps oneself. This might happen in the most direct way as when, for example, two organisms band together, united they conquer, divided they fall, or when one organism helps another anticipating benefits in return (e.g., cleaner fish get fed, and the cleaned fish are freed from parasites (Trivers, 1971); or cooperation might occur in more subtle ways, as when organisms are related and thus share the same genes, that is, have identical copies of the same alleles. What matters from an evolutionary perspective is that a gene gets passed on. If it does so better than others, no matter how, it will obviously increase its representation in future generations. It might, therefore, in certain circumstances pay an organism to help a close relative (i.e., sharer of many genes) to reproduce, even at the expense of its own reproduction to the point of sterility. Reproduction by proxy, as it were! (Hamilton, 1964a, b).

These two mechanisms for promoting altruism at the behavioural level, the one expecting direct returns and possibly operating between non-relatives and the other involving indirect reproduction through relatives, have been labelled 'reciprocal altruism' and 'kin selection' respectively. (For more details, see Ruse, 1979b.) I do not think it too much to say that kin selection, in particular, has been the basis for one of the triumphs of sociobiological speculation, if not to say indeed one of the triumphs of evolutionary speculation. There has long been a question about the reason for the sterile cases in Hymenoptera (ants, bees, and wasps). Why do some insects devote their whole lives to the welfare of their fertile nest-mates? With one blow, thanks to kin selection, sociobiologists have been able to solve the mystery. Hymenoptera have a funny sexual system. Females are like humans (male and female): they are diploid, that is to say, they have two half sets of chromosomes, one half set from each parent. Males however are haploid: they have only one half set of chromosomes, those from their mothers. They have no fathers, and indeed inseminated fertile females can control the sex of their offspring by fertilizing or not fertilizing one of their own eggs. But this haploid-diploid sexual system, which no doubt had good evolutionary causes, has an interesting consequence. Females are more closely related to sisters than to daughters! Sisters share the same paternal genes ($\frac{1}{2}$ of the set) and $\frac{1}{2}$ of the same maternal genes ($\frac{1}{2} \times \frac{1}{2}$ of the set). This means that sisters share $\frac{3}{4}$ of the same genes, whereas mothers and daughters share only $\frac{1}{2}$ of the same genes. As a consequence, for kin selection reasons, it is in a female's biological interests to forego her own reproduction and help to raise fertile sisters! Hence, sterile castes. Males, to the contrary, are not more closely related to siblings than to daughters (they have no sons) and expectedly we do not find male sterile worker Hymenoptera. (For further details see Wilson, 1975a, Chapters 5 and 20; also Oster and Wilson, 1972).[2]

The reader now has a taste of animal sociobiology, and I hope some flavour of its excitment. Even now in this essay, pressing philosophical questions are bubbling up, not the least of which devolve on all of this talk about 'altruism', 'selfishness', 'enlightened self-interest', and the like. For the moment, let us shelve these questions: they will not be forgotten. Here let us turn to the matter which caused all the controversy, namely Wilson's treatment of our own species, *Homo sapiens*.

8.2. HUMANS AS SEEN THROUGH THE LENS OF SOCIOBIOLOGY

Humans are surely distinguishable from all other organisms by their culture and by their sophisticated social environment. Moreover: "It is part of the conventional wisdom", notes Wilson, "that virtually all cultural variation is phenotypic rather than genetic in origin" (Wilson, 1975a, p. 550). And in support of such a position, Wilson allows that vast cultural changes occur at rates far too high to put down to a direct function of genetic change. Hence, it would seem that humans no longer fall within the domain of sociobiological inquiry. Nevertheless, argues Wilson: "Although the genes have given away most of their sovereignty, they maintain a certain amount of influence in at least the behavioural qualities that underlie variations between cultures. Moderately high heritability has been documented in introversion-extroversion measures, personal tempo, psychomotor and sports activities, neuroticism, dominance, depression, and the tendency towards certain forms of mental illness such as schizophrenia . . . Even a small portion of this variance invested in population differences might predispose societies toward cultural differences" (ibid, p. 550). It seems fairly obvious that what Wilson has in mind at this point, is some kind of ripple effect where relatively minor genetic differences could explode up into major behavioural differences. The schizophrenic Joan of Arc hears voices which leads her to inspire the French, not only against the enemy of her day, but even down to our own time when de Gaulle could rally his countrypeople by invoking her memory (Ruse, 1979b, p. 78).

In arguing that cultural differences might be in part genetic, this is not to say that Wilson is claiming that only differences (as opposed to cultural similarities) would be genetic. Far from it. By a comparative taxonomic study between primates, Wilson points to some fairly common human characteristics also shared generally by primates. Just as cultural variations might be due to genetic differences, so apparently these cultural constants might be due to genetic similarities (between all humans and also with primates). Amongst the constants Wilson finds "aggressive dominance systems, with males generally dominant over females; scaling in the intensity of responses, especially during aggressive interactions; intensive and prolonged maternal care, with a pronounced degree of socialization in the young; and matrilineal social organization" (Wilson, 1975a, p. 551).

The general position staked out, Wilson now steers his discussion into more particular topics. Thus, for example, there is an analysis of role playing in human societies and its possible genetic underpinnings. Is it possibly the case that we are what we are in a society and that we do what we do in a society because of our genes? Although Wilson suggests that theoretically (i.e., population genetically theoretically) we could get a kind of clustering of genes reflecting different roles, generally speaking, he feels that this line of approach may not be very fruitful. For instance, the 2000-year-old caste system in India seems to have brought about, or reflects, hardly any differences. "Even so," continues Wilson, "the influence of genetic factors toward the assumption of certain *broad* roles cannot be discounted" (ibid., p. 555, his italics). In support of this suspicion, the reader is invited to consider the topic of male homosexuality. Possibly this has a genetic basis. It could be, since *prima facie* homosexuality is a genetic unfitness because homosexuals seem less likely to leave offspring, that genes for homosexuality are maintained in populations because of superior heterozygote fitness (i.e., whilst the homozygote homosexuals are less fit, their heterozygote relatives are superfit). Alternatively, it could be that homosexuals are in the position of worker ants! "Freed from the special obligations of parental duties, they [i.e., homosexuals] could have operated with special efficiency in assisting close relatives. Genes favoring homosexuality could then be sustained at a high equilibrium level by kin selection alone" (ibid., p. 555). In *On Human*

Nature, Wilson has developed this theme in somewhat more detail. Later in this essay I shall return to this.

To me, as a philosopher, some of Wilson's most interesting comments come when he touches on religion and philosophy. He sees culture, generally speaking, as adaptive ("It is useful to hypothesize that cultural details are for the most part adaptive in a Darwinian sense" ibid., p. 560), and this applies obviously both to religious and philosophical beliefs. Thus: "It is a reasonable hypothesis that magic and totemism constituted direct adaptations to the environment and preceded formal religion in social evolution" (ibid., p. 560). Indeed, Wilson goes as far as to quote, with what I think is approval, fellow-traveller Robin Fox (1971), to the effect that, were some humans raised in isolation, eventually they or their descendents would develop religious beliefs and practices – and this despite the fact that (in Wilson's opinion) so much of religion is false! "Men would rather believe than know . . ." (Wilson, 1975a, p. 561). Ethics also, Wilson ties to the genes. Not surprisingly, he sees ethics as coming out of the selection for selfishness and altruism, and he points out that selection may well promote different interests for different people, even for different age-groups. Hence, true ethical understanding presupposes sociobiological understanding (ibid., p. 564).

Rushing along, we find that Wilson discusses such topics as territoriality and tribalism, and warfare. Regarding the latter, he refers to some authors, who "envision some of the 'noblest' traits of mankind, including team play, altruism, partriotism, bravery on the field of battle, and so forth, as the genetic product of warfare" (ibid., p. 573). And so finally Wilson comes to an end – not, I might add, on a particularly happy note. We are soon going to have to plan our society in a pretty drastic way. Who knows but that, as we strive to eliminate unwanted human tendencies, we might find them linked genetically (i.e., through 'pleiotropic' genes) to desirable qualities. As we strive to throw out the unwanted bathwater, the genetically linked baby may go too!

8.3. OTHER SOCIOBIOLOGICAL CLAIMS

In concluding this preliminary exposition of sociobiology, I reacknowledge that I have, perhaps rather unfairly, been restricting my discussion to one work by one person. Although as a formal, integrated discipline – inasmuch as it is one – sociobiology is young, it is by no means the work of an un-aided E. O. Wilson. This was not so in 1975 and it is even less so today. Two who perhaps deserve special mention, particularly for their work in human sociobiology, are Robert L. Trivers and Richard D. Alexander. The former, particularly, has made some of the most audacious or, depending on one's viewpoint, foolhardy claims to come from the sociobiological stable. For example, Trivers has not been beyond suggesting that squabbles between parents and children over bedtimes might well be a function of the genes — ". . . when parent and child disagree over when the child should go to sleep, one expects in general the parent to favor early bedtime, since the parent anticipates that this will decrease the offspring's demands on parental resources the following day" (1974, p. 260; see also Trivers, 1971, 1972). Alexander, perhaps more than other sociobiologists, has combed the social science literature looking for human behavioural backing for genetic speculations (Alexander, 1971, 1974, 1975, 1979). He suggests, for instance, that the so-called 'mother's brother' phenomenon can be given sociobiological backing. In some societies the adult male responsible for child care is not the father but the mother's brother. *Prima facie*, this seems to violate natural selection, because fathers and offspring are more closely related than uncles and nephews and uncles and nieces. However, argues Alexander, if paternity in a society is often in doubt, mother's brother care makes good genetic sense because one is in virtually no doubt as to one's sister's children (Alexander, 1979). Similarly, Alexander argues that the almost-universal human taboo on incest can be explained in terms of genes and selection. Incest is biologically deleterious because inbreeding from incest makes for homozygosity of alleles of a kind which have dangerous effects only when paired with like mates. Hence, there has been strong selective pressure promoting a genetic basis for behavioural avoidance of intercourse with close relatives. And in addition, infanticide is so explained in terms of genes. Alexander (1975) points out, for example, that in some primitive societies in times of need or hardship, babies are killed better to protect older offspring — even to feed the older offspring. It is argued that this makes good genetic sense from the parent's viewpoint: better one offspring than none.

Obviously, in this kind of explanation of Alexander, we have a direct parallel with explanations of similar practices in insects, where some eggs (so-called 'trophic' eggs) are regularly fed to other offspring. And with this point — one which re-emphasizes the way in which sociobiology really tries to view human beings in the same evolutionary way as it views the rest of the animal world — we have an appropriate place to stop the exposition and turn to critical responses to sociobiology. As the criticisms are expounded I shall try to evaluate the pros and cons of the case for human sociobiology.

8.4. IS HUMAN SOCIOBIOLOGY FACIST?

The chief critics of sociobiology have been members of the 'Science for the People' group of radical scientists, members of which group we have already encountered in the attack on recombinant DNA research (see, especially Allen *et al.*, 1975, 1976, 1977). A major, probably *the* major, objection that these critics have of human sociobiology is that it is not genuine science but right-wing ideology dressed up to look like science, and that consequently the enterprise reveals and 'justifies' all sorts of prejudices against different types of people. In order to make this case the SftP critics begin by placing sociobiology firmly in the tradition of biological determinism, a science-cum-philosophy which "attempts to show that the present states of human societies are the result of biological forces and the biological 'nature' of the human species" (Allen, *et al.*, 1977). This all seems fairly true, although one might wonder how any scientific theory of humans could not, in some sense, be biologically deterministic, even the most environmentalist, given the way that biologists perpetually emphasize how organisms are a product both of their genes and of the environment. Of course, what the critics intend to show is how sociobiology relates humans firmly to their genes in a way that environmentalists do not, and in this they are obviously right — sociobiology is nothing, if not this. The average environmentalist notes various biological needs like hunger and then ignores them as much as possible, at least insofar as such needs might have a biological function. The sociobiologist virtually starts at this point.

But obviously, the SftP critics have more in mind than merely a taxonomy of biological disciplines, separating out those which are really biologically deterministic from those which are not. Such determinist theories, we learn, reflect 'socio-economic prejudices'; they are apologetics for the *status quo*; and indeed they lead to, certainly support, the philosophy of the gas-chamber.

For more than a century, the idea that human social behaviour is determined by evolutionary imperatives and constrained by innate or inherited predispositions has been advanced as an ostensible justification for particular social policies. Determinist theories have been seized upon and widely entertained not so much for their alleged correspondence to reality, but for their more obvious political value, their value as a kind of social excuse for what exists. (Allen *et al.*, 1977; see also Allen *et al.*, 1976, p. 182.)

And the critics flesh out their case by citing the instance of Konard Lorenz, who began with geese and ended by echoing Hitler's racial policies of extermination.

Now, considering matters at a general level, what can we say about things?

Is it true that human sociobiology is little more than a neo-facist wolf in a
scientific lamb's clothing, and that consequently anyone who has any part
in it is tainted with the same filthy beliefs that lead to the extermination of
six million Jews? Let me make my own position categorically clear. I believe
that such a charge is false, and indeed that it is wrongheaded to the point of
cruelty. Wilson is no racist. There is absolutely no hint in his work, or that
of fellow human sociobiologists, that any claims about genetic differences
between human groups might be the thin end of the wedge for claims about
the inferiority of some types, within or crossing national boundaries. Indeed,
let me say to the contrary, what impresses me most strongly is the extent to
which human sociobiology affirms the unity of humankind. The differences
between, say, the hunter in the Kalahari desert and the New York business
executive are too obvious to be really worthy of note. But what the sociobiologist argues is that much that motivates the hunter is identical to much
that motivates the executive. They are both after the kinds of things which
lead to success in the evolutionary struggle and they are after them for the
same sorts of reasons. For instance, just as status aids the hunter, so status
aids the executive. Just as the hunter seeks a mate and cares for children for
evolutionary reasons, so also does the executive. Just as the hunter behaves
in certain ways towards fellows (e.g., lies sometimes, cheats sometimes, is
friendly sometimes) because these are things which have proven their evolutionary worth in the past, so the executive behaves in the same way towards
fellows because of the same evolutionary worth. One may, of course, want to
deny that one can tie down behaviour like cheating to evolutionary foundations. The point here, however, is that the human sociobiologist does see such
links, and that the human sociobiologist wants to universalize these across
humanity. This, I believe, goes far to defend the sociobiologist against charges
that he/she picks out certain groups as different and possibly inferior.

There is another way in which one might fear that human sociobiology,
explicitly or implicitly, supports an incredibly reactionary view of humankind. This is a fear which has been articulated by one of the many social
scientist critics of sociobiology, the eminent anthropologist Marshall Sahlins
(1976). Even if one agrees that sociobiology does not pick out certain peoples
or races as inferior *per se*, in another sense perhaps it supports anyone who
wants to treat such peoples or races as inferior. The reason for this is that
sociobiology sees life as a big battle for existence, with each and every one
of us selfishly out for what we can get. Even friendship is no more than
enlightened self-interest. There is no genuine altruism. Hence, if it is in our
interests to put down certain people — keeping blacks at the low end of the

socio-economic scale so that we have a ready supply of janitors and the like — then perhaps sociobiology gives justification for such actions. If this is not neo-facism, then nothing is.

I must admit that, with respect to this point, in some ways sociobiologists have been their own worst enemies. They bandy about terms like 'struggle' and 'self-interest' with nary a hint of awareness that such language may rebound on them. Nor does it much help their case when they publish books with titles like *The Selfish Gene* — a popularization of sociobiology which, I hasten to add, I enjoyed immensely (Dawkins, 1976). However a little thought soon shows that, for all the language, sociobiologists are not really guilty of such henious sins as a critic like Sahlins charges. Obviously, when the socio-biologist says that certain actions are 'selfish' or a function of 'self-interest', the terms are being used metaphorically, both in the animal *and* the human world. Suppose, for instance, a sociobiologist explains certain actions or customs as a function of kin selection. A good example was provided earlier, namely Richard Alexander's (1979) kin selection explanation of the custom in certain societies whereby the adult male responsible for care of children of any particular woman is not the husband but the woman's brother(s): where paternity is often in doubt, although an uncle is less closely related than a biological father, from an evolutionary perspective, altruism towards un-doubted relatives is preferable to altruism towards those who may not be related to one at all (see also van den Berghe, 1979).

In a sense, one might say that what occurs here in the mother's brother situation is selfish behaviour or something furthering self-interest. But clearly one does not necessarily mean this in a literal sense. A selfish person is one who deliberately does something for their own ends: 'with malice afore-thought'. If I take all the butter and leave none for anyone else, that is self-ishness. However, there is no implication at all that the mother's brothers are deliberately furthering their own evolutionary ends, any more than there is implication that the sterile worker ant selfishly works out the best genetic strategy for herself. Indeed, Alexander is explicit in his claim that at an inten-tional level, the selfless motives might be entirely genuine: that things may be working in one way down at a genetic causal level does not at all deny that as far as the individual human is concerned, altruistic actions and desires are quite sincere. This is not to say that selfishness never occurs; of course it does. But it is to say that a sociobiological perspective of human beings does not necessarily imply an unrestrained scenario(!) of Nature red in tooth and claw. There is no endorsement of John D. Rockefeller's view of the business world, when he told a Sunday school class that it was right and proper that

Standard Oil should push other businesses to the wall because this was the
way Nature intended (Hofstadter, 1959).

On general grounds therefore, I would argue that the case for claiming that
human sociobiology is politically reactionary fails. Moreover, I think one can
see ways in which sociobiology might be used for good ends, whether these
he judged by utilitarian or Kantian criteria. For instance, Wilson suggests that
xenophobia (fear of strangers) may be genetic: "Part of man's problem is that
his inter-group responses are still crude and primitive, and inadequate for the
extended relationships that civilization has thrust upon him . . . xenophobia
becomes a political virtue" (Wilson, 1975a, p. 575). Surely a fuller understanding of this problem may help immeasurably in reducing intergroup
tension, such as racial strife, as well as the frictions between nations? To
argue otherwise seems akin to claiming that the search for the cause of cancer
has no relevance to a search for the cure.

However, the critics of human sociobiology will still not be satisfied by my
reassurances. They will feel that when we look at specific groups and at what
sociobiology has to say about them, we see that prejudice continues to rear
its ugly head. One group particularly that the SftP critics single out as having
been unfairly judged by the sociobiologists, is women. It is argued that
human sociobiology is rankly sexist, valuing males and things male to an
inordinately great extent. Since this is so important a criticism and raises such
significant questions about the whole nature of science, not simply sociobiology, I shall consider it in its own right in a separate essay, the next in this
collection. Here, let me try to defuse the critic's objections by considering
the sociobiological treatment of another group where one might feel that
bias come through masked under the guise of objective science. I refer to
something which has already been mentioned, namely the sociobiological
treatment of human homosexuality.

8.5. IS SOCIOBIOLOGY PREJUDICED AGAINST HOMOSEXUALS?

As we have seen, there are at least two explanations of human homosexuality
proferred by the sociobiologists. On the one hand, it is argued that possibly
homosexuality is a function of balanced heterozygote fitness, where less-fit
homozygotes for 'homosexual' genes are balanced out by their super-fit
heterozygote, heterosexual siblings. (This model was first suggested by
Hutchinson (1959).) On the other hand, it is surmized that possibly homosexuality is a function of kin selection. About this latter possibility, Wilson
has recently written as follows:

How can genes predisposing their carriers toward homosexuality spread through the population if homosexuals have no children? One answer is that their close relatives could have had more children as a result of their presence. The homosexual members of primitive societies could have helped members of the same sex, either while hunting and gathering or in more domestic occupations at the dwelling sites. Freed from the special obligations of parental duties, they would have been in a position to operate with special efficiency in assisting close relatives. They might further have taken the roles of seers, shamans, artists, and keepers of tribal knowledge. If the relatives — sisters, brothers, nieces, nephews, and others — were benefitted by higher survival and reproduction rates, the genes these individuals shared with the homosexual specialists would have increased at the expense of alternative genes. Invariably, some of these genes would have been those that predisposed individuals toward homosexuality. A minority of the population would consequently always have the potential for developing homophilic preferences. Thus it is possible for homosexual genes to proliferate through collateral lines of descent, even if the homosexuals themselves do not have children. This conception can be called the 'kin selection hypothesis' of the origin of homosexuality. (Wilson, 1978, p. 144—5.)

On the basis of explanations like these, does sociobiology stand convicted of bias against homosexuals? I would suggest not, although it must be allowed that the sociobiologists have not always taken a quite so strong and unambiguous stand as they might have done. Indeed, at one point, Wilson actually refers to homosexuality as an 'abnormality' and classifies it along with other human aberrations like cannibalism! (Wilson, 1975a, p. 255). Not even right-wing Church people usually put homosexuality quite this low, although admittedly St Thomas Aquinas (1968) did rate homosexuality below rape: homosexuality is a violation against God, whereas rape is merely a violation against humans. However, if one looks at the context of Wilson's discussion one see that, in fact, what he is referring to is heterosexuals behaving homosexually when under strange conditions, at times of great stress or when in prison. This surely is abnormal behaviour, just as it would be were homosexuals forced to act heterosexually. The situation is analogous to a man being made to wear a skirt. Such behaviour is not abnormal for a woman but it is for a man. One could, therefore, wish that Wilson had been a little more careful in his language, but sociobiology itself does not stand convicted because of such sloppy use of terms.

However, one might feel that prejudice is actually built into the very models sociobiologists put forward to explain homosexuality. In particular, one learns from the balance model that homosexuals are 'unfit', and one learns from the kin selection model that basically, homosexuals are making the best of a bad job as they try to reproduce vicariously rather than directly. If all of this is not to view homosexuals as second-rate, then nothing is. But by now it should be obvious that this kind of criticism fails. It is certainly true that the balance model assumes homosexuals to be unfit in the sense that they will have fewer offspring than heterosexuals, but this does not mean that they are unfit in any physical sense. The mule is sterile, that is it is totally unfit in a biological sense, and yet it is far tougher than either parent. Similarly, the homosexual could be just as tough, or happy, or whatever, as the heterosexual relatives. As far as the kin selection model goes, again there is no direct implication that the homosexual will be inferior. To use a metaphor, natural selection is totally indifferent to the means it uses to gain its ends. Perhaps, indeed, someone may become a homosexual because physically they are too weak to make it as a heterosexual. However, it could equally be that someone becomes a homosexual because they have prized characteristics not possessed by heterosexuals. It has been suggested, for instance, that homosexuals might have a higher IQ than heterosexuals (Weinrich, 1978). It could be that such a characteristic makes someone a better altruist and thus one can reproduce more efficiently by proxy, even though potentially one would be quite capable of normal (i.e., average) reproduction, were one a heterosexual. In short, there is nothing in either model which invariably labels homosexuals as worse than heterosexuals.

Let me try one more attack on behalf of the critics. It might be argued that, whatever the nature of the explanatory models, sociobiology shows prejudice against homosexuals in the very attempt to explain homosexuality! In picking out homosexuality as something 'needing' explanation, we put homosexuality alongside with earthquakes, gonorrhea, and inflation, which are other things which all 'need' explanation. If sociobiology really had no bias against homosexuality, argues the critic, then it would not look upon it as a problem in the first place. Sociobiology treats homosexuals as 'queer' (pun intended).

In respects, I am sympathetic to this criticism. I agree with the underlying assumption that science is not some neutral, objective enterprise, which simply reports in a disinterested way on what is 'out there'. As I shall explain in more detail in the next essay, I believe science can show values and interests, just like any other human endeavour. Nevertheless, the sociobiologists can be defended against charges of bias at this point. One may or may not accept Darwinian evolutionary theory, but the crucial point is that sociobiologists do accept it. We have seen that their whole programme is no more than an attempt to explain social behaviour within the Darwinian framework. But this means that sociobiologists accept the central premise that life is a struggle for existence, or more pertinently, a struggle for *reproduction*. For them, therefore, homosexuality is a problem: not a moral problem, but a conceptual one. Homosexuality is behaviour which, *prima facie*, reduces reproduction — hence as people committed to the belief that (in one sense) reproduction is the key to everything, sociobiologists would be derelict were they not to attempt to explain homosexuality. Their explanations may or may not be adequate. But that is another matter. What is certain is that sociobiologists show no prejudice in attempting to explain homosexuality. (Having combed the literature I have discovered hints towards at least two more sociobiological models for homosexuality. I discuss them in Ruse (1981), and defend them against charges of bias.)

8.6. THE TESTABILITY OF SOCIOBIOLOGY

Having defended human sociobiology against charges of moral bias, let us turn now to criticism of a rather different kind that the SftP group have levelled against it. At what one might call the 'epistemological' level there are two major criticisms that the critics make of Wilson's position. On the one hand, they argue that sociobiology, especially human sociobiology, is not open to empirical test. Therefore, since it is unfalsifiable, it is metaphysical, and hence only a pseudoscience. On the other hand, they argue, perhaps not entirely consistently, that sociobiology is false. We will consider these two charges in turn.

The main charge of unfalsifiability comes over the question of adaptation. The critics charge that sociobiology claims that all behaviours are or were adaptive, and that it is so constructed that this claim will always be found true. They write:

When we examine carefully the manner in which sociobiology pretends to explain all behaviours as adaptive, it becomes obvious that the theory is so constructed that *no tests are possible*. There exists no imaginable situation which cannot be explained; it is *necessarily confirmed by every observation*. The mode of explanation involves three possible levels of the operation of natural selection: 1. classical individual selection to account for obviously self-serving behaviors; 2. kin selection to account for altruistic or submissive acts towards relatives; 3. reciprocal altruism to account for altruistic behaviors directed toward unrelated persons. All that remains is to make up a 'just-so' story of adaptation with the appropriate form of selection acting. (Allen *et al.*, 1977, p. 24. See also Allen *et al.*, 1976, p. 185.)

We must tread carefully here. First, we must distinguish between on the one hand, a theory or hypothesis being falsifiable, and, on the other hand, its being falsified. It is obviously not to the discredit of sociobiology if there do not exist in the world any phenomena which are known to show it false. Suppose, for example, sociobiology based its whole case on individual selection without invoking such things as kin selection. Then, humans quite apart, we would know that sociobiology could have only very limited application (i.e., claims about its universal application would be false), for many members of the insect world would stand in living refutation of it. So, in itself, the fact that sociobiology has answers is no fault.

But this leads straight to the second point. One might agree that, as it stands, sociobiology has answers, but feel that it has no right to stand as it does. In particular, one might feel that all this talk of 'kin selection' and 'reciprocal altruism' is just so much *ad hoc* padding to save central premises which one is refusing to put to the test of experience, or, as the critics rather imply, that these new variants of selection are just invented so that sociobiology can be applied to humans without falsification. However, this seems not to be true. Kin selection is indeed fully accepted by the critics themselves! "There does exist data supporting the idea that kin selection is effective for some traits in social insects . . ." (Allen *et al.*, 1977, p. 26). And although the critics seem rather less enthused about reciprocal altruism, these self-same critics bring no argument against it. Moreover, kin selection has been developed from good populational genetical premises and, quite apart from humans, seems needed and confirmed in the animal world. We ourselves have seen its brilliant and incisive application to the Hymenoptera, and its scope is presently being extended right up the animal chain. Hence, that the sociobiologists appeal to these various kinds of selection seem unobjectionable. They are not just invented to explain peculiarities of humans. (For an experimental test of kin selection see Trivers and Hare (1976); but see also Alexander and Sherman (1977).)

At this point thought, it might thirdly be charged that sociobiology is still objectionable because for all the legitimacy of its various kinds of selection, it contains at its core an objectionable unfalsifiable premise, namely that all features of the organic world — particularly all facets of human behaviour — are, or were, adaptive (Lewontin, 1977). Nothing can attack this claim. But, of course, this criticism is just not based on true premises. As we have seen, Wilson entertains the hypothesis that homosexuality in itself is not adaptive. Perhaps the homosexual has two homosexual genes homozygously — genes which are kept in the population only because the heterozygotes have superior, reproductive fitness. It is indeed true that sociobiologists look at all behaviour with an eye to its adaptive value. But in this they are no different from other Darwinian evolutionists. It is, after all, a sensible heuristic method-ological approach — many characteristics once thought to be without adaptive value, such as the banding on snails, are now thought adaptive (Sheppard, 1975, p. 129). The paradox is that it is Richard Lewontin who has most insistently criticized sociobiology over this adaptivist viewpoint, and yet when he is not on his war-horse against sociobiology, it is he who has most brilliantly defended the whole Darwinian outlook (in Lewontin, 1978)! Apparently philosophers are not the only ones in this world who want to have things both ways at once. Could there be a sociobiological explanation?

One can go on almost indefinitely considering possible charges that socio-biology is unfalsifiable. But as a fourth and final suggested charge, it might be claimed that, although sociobiology does allow non-adaptive features, it is unfalsifiable in the sense that whatever phenomenon is turned up, it feels that there is some explanation following from or compatible with its list of alternatives, adaptive and non-adaptive. There is, in short, nothing that would make the sociobiologist say: 'I'm wrong'.

Now, in reply, metaphysical though this practice may be, I am not sure *a priori* that, for all that Sir Karl Popper (1959) says otherwise, as a practice, this position is so very objectionable — at least that speaking generally, it is non-scientific. I am going to require an awful lot of evidence to persuade me that the earth is not round or that astrology is not totally fraudulent. But, in any case, if with some justification one points out that the claims of socio-biology are hardly in the same camp as claims about the roundness of the earth, I wonder if this charge is indeed true? Suppose for example, anthro-pological or historical evidence were uncovered showing that certain tribes openly and deliberately practiced incest for generation after generation — that if (as a male) one did not mate up with one's mother, sister, or daughter, one was looked upon as a social pariah. This, it seems to me, would call for substantial rethinking by sociobiologists. Or suppose one came across a society where type A's gave freely to type B's, to their own detriment, with-out any rewards at all. There would not be much reciprocation about this altruism. Of course, one might claim that some sociobiologists would wriggle to get round even something like this, but I wonder if we have not got to the point where we should then be blaming sociobiologists not sociobiology. That some people will not let a theory be falsified does not mean that the theory itself is unfalsifiable — as much as any theory is that is (Kuhn, 1962).

Incidentally, in reply to an obvious objection that will be made at this point, I find myself singularly unimpressed by such supposed counter-exam-ples to incest taboos as the Pharaohs. If these are the best that the critics can come up with, then their case is thin indeed. Such counters are absolutely minute compared to the massive counter-evidence that most laws of physics have to live with — how many Egyptians at that time never married their sisters, let alone how many Egyptians of all time, or Arabs of all time, or humans of all time? Rather than throwing out universal incest taboos, it seems far more reasonable to explain away the very rare exceptions, as indeed sociobiologists have attempted to do:

Dynastic incest restricted the number of legitimate claimants on the most important resource in those societies, supreme political power. The possible reduction in biological fitness of one's children-nephews was more than made up for by the monopolistic retention of extraordinary resources, especially when those resources gave one access to innumerable lesser wives and concubines who, although they could not bear future kings, could certainly bear children. (van den Berghe, 1979, p. 78.)

So far, in considering the charge of unfalsifiability, I have been defending sociobiology. Before concluding this discussion I must register a *caveat* — one which may perhaps give much pleasure to the critics of sociobiology. I have been arguing that a case can be made for the falsifiability of sociobiology. But this, of course, is a long way from saying that it is a well-confirmed theory, which obviously, as it applied to humans, it is not. Take, for example, the question of homosexuality. Wilson's speculations about the genetic causes of homosexuality are really little more than that — speculations. In the first place, he gives no real proof that homosexuality significantly limits reproductive fitness. *Prima facie*, it certainly seems true that the homosexual has less children than the heterosexual; but does he/she really? It is, in fact, true that a recent study done by the Kinsey Institute in San Francisco implies that homosexuals are less likely to marry than heterosexuals, and that if they do marry, they have far fewer children (Bell and Weinberg, 1978). So perhaps for our own society, reduced reproduction by homosexuals is a reasonable inference. (For more details see the final essay.) But how typical are San Francisco homosexuals? In many societies one has to conceal homosexuality. Do homosexuals under these conditions, conceal their lot and breed. And what about lesbians? If it is the same set of genes controlling sexual preference at work for men and women, then possibly these genes (whatever their nature) have been kept at high levels in populations because, until recently at least, no one very much consulted women about their sexual preferences, and so the genes have been maintained and passed on via the distaff side of the human race.

But, all of this hypothesizing — falsifiable perhaps but in no way confirmed

— leads on to the second point. Hardly any evidence is offered that homo-sexuality is directly under the control of genes. Certainly, if the balance hypothesis be true, a necessary (although not sufficient) condition is that identical twins share the same sexual orientation. In fact, there are reports of undoubted identical twins with different sex preferences, so clearly the balance hypothesis cannot be universally true — it is not simply falsifiable but falsified! (Rainer *et al.*, 1960; Green, 1974). However, whether from a general perspective twins tend to have the same preferences, suggesting that the balance hypothesis could hold in part, is really not checked. It is true that Wilson refers to some sweeping claims from the past, claims that twins do invariably have the same sexual orientation, but these are based on such questionable methodology that at this point one has to conclude that, as the Scots say, the case is 'not proven' (see Kallmann, 1952; Heston and Shields, 1968).

In fairness, one must note that one indefatigable researcher, James Weinrich, has combed the anthropological literature trying to find empirical under-pinning for the kin selection hypothesis for homosexuality (Weinrich, 1976). In preliterate societies, do we find that homosexuals aid their kin? Do we find reasons why such people might have opted (not necessarily consciously!) for the alternative homosexual strategy? Do we find that there are exceptions which, as it were, prove the rule? Moreover, Weinrich feels that his search has not been entirely without results. For instance, there is good evidence sup-porting the claim that homosexuals frequently do take on roles which would enable them to aid close relatives significantly in terms of prestige and mate-rial goods. Nevertheless, the case is hardly yet made in a definitive fashion. And this, I would suggest, makes reasonable my general feeling about socio-biology (*qua* humans) as it stands at present. We find absolutely fascinating hypotheses; but to date basically just that — hypotheses, not theories. (See Ruse (1981) for much more discussion of the sociobiology of homosexuality, including a full assessment of its truth-status.)

For further evidence supporting my feelings about the present status of human sociobiology, let us turn to the other major charge that the critics level against sociobiology, namely that it is false.

8.7. THE FALSITY OF SOCIOBIOLOGY

The critics argue that, although essentially sociobiology is unfalsifiable: "There does exist, however, one possibility of tests of such [sociobiological] hypotheses, where they make specific *quantitative* predictions about rates of change of characters in time and about the degree of differentiation between populations of a species" (Allen *et al.*, 1977, their italics, p. 27). In particular, argue the critics, if sociobiology be true, then (restricting our gaze to the human world) we should find major cultural changes are accompanied by (since they are a function of!) significant genetic changes. Moreover, we should find significant genetic differences between populations, again reflecting (and causing) major cultural differences.

But, neither of these predictions are true. Temporally speaking, major cultural changes have occurred in a period of time far too rapid to have been caused by genetic changes — at least changes as allowed by orthodox population genetical theory, which, as we have seen, Wilson accepts. Thus, for example, the rise and fall of Islam took less than 30 generations, and so that cannot have been directly fired by the genes. Co-temporally speaking, we find that there just is not the genetic variation between populations that would explain the vast cultural differences. ". . . we know from studies of enzyme specifying genes that at least 85% of that kind of human variation lies *within* any local population or nation, with a maximum of about 8% between nations and 7% between major races" (Allen *et al.*, 1977, their italics, p. 28).

Of course, the sociobiologists, Wilson in particular, are not ignorant of these facts. Indeed, they acknowledge their like explicitly. What they argue is that, even within these limitations, there is room for the approach they would take towards human beings. Thus, nothing said by the critics denies that there are behavioural differences between humans in any population. Some are heterosexual; some are homosexual. Some are schizophrenic; some are not. Some are extroverts; some are introverts. And so on. Possibly, indeed there is good evidence that very probably, some of these differences are a fairly direct function of the genes. There are genetic differences within populations and studies show that some behavioural characteristics seem to be heritable. Schizophrenia is a case in point. A propensity towards alcoholism has also been suggested, as have other things, such as certain kinds of criminal behaviour (see McClearn and De Fries, 1973; Levitan and Montagu, 1977). Of course, in this context one thinks of questions to do with intelligence. Without wanting to muddy further already-dirty water, one can say that certain gross intelligence deficiencies do seem due to gene or chromosome abnormalities (Down's syndrome, for instance); that in talking of "intelligence", one is bundling together a number of abilities which ought to be separated; and that whatever else may be the case, the environment is crucially important in the development of intelligence (Ruse, 1979b). But, be this as it may, it does seem that the study of the genetic bases of individual behaviours is a viable, probably fruitful, programme. I would add, however, that I am not sure about how much sociobiology itself has added to this enterprise so far. As I have pointed out, the sociobiologists' speculations about (say) the causes of homosexuality are still, to a large extent, in the realm of unproven hypothesis. What solid results we have to date seem to come from regular behaviour geneticists. Indeed, were I one of them, I might be feeling just a little put out by all the publicity that the sociobiologists are getting.

In the apportioning of due credit on one side, however, we can agree that sociobiologists might still profitably explore intra-group differences. It seems unlikely that all positive results will prove false. In the fashion, sociobiologists might (and do) look for common patterns between populations — even between populations which, in most respects, have significant cultural differences. Then, having identified such patterns, they might look for genetic bases to them. Again, however, one must stress that as things stand at the moment, although we are not dealing with the *a priori* false, the search for genetically based common patterns is more of a sociobiological programme than sociobiological achievement. First one needs, as one generally does not have at present, clear specification of what these common patterns are, together with full anthropological documentation. Then one needs evidence that their basis is genetic. At present it is not axiomatic that even some of the strongest patterns are genetically based. Take incest avoidance. Apparently the effects on humans of close inbreeding are devastating (Adams and Neel, 1967). Could it not be that after a certain level of intelligence development, humans realized its dreadful effects and therefore decided (i.e., decided culturally) to stop it? Perhaps this is not so; but it does show that the socio-biological case must be argued for, not assumed.

But in fairness, I must not end this section on a totally negative note. As far as incest avoidance is concerned, sociobiologists do and can properly refer to the fascinating findings from the Israeli kibbutz (Shepher, 1971; 1979; Tiger and Shepher, 1975; van den Berghe, 1979). Children of the kibbutz are raised together, non-relatives with non-relatives. Although there are absolutely no formal sanctions against them, adults find it virtually impossible, from a psychological standpoint, to have sexual relations with those with whom they were raised. Could it not be that incest taboos are underscored by a kind of antisexual imprinting, which occurs between young children raised together, which potential for imprinting has a genetic base fashioned by selection for incest avoidance? If this were the case, then human sociobiology could claim a strong piece of positive evidence.

   In the same line I must note that Alexander has gathered impressive evidence for his mother's brother's hypothesis, showing that, generally, the phenomenon occurs in all and only those societies where paternity is in doubt (Alexander, 1979). Time and again one finds that, in societies where there is much extra-marital sexual intercourse, it is the mother's brother who is responsible for child care and attention. I am still not yet convinced that one can properly treat human sociobiology as an absolutely proven theory in any significant respect, but it is surely unfair simply to discuss it as false or trivial. Alexander, for instance, seems to progress in exactly the same path of successful prediction and confirmation as do scientists in more orthodox fields. The critics may truly point out the limitations circumscribing human sociobiology, but they have not, as yet, shown it totally untrue. If anything, the tide seems to be flowing the other way, somewhat.

8.8. SOCIOBIOLOGY AND PHILOSOPHY

I suspect that my rather hesitant attitude about their subject will not really dismay the sociobiologists. Perhaps my strongest impression from reading *Sociobiology: The New Synthesis* is how very little we actually know as yet about animal behaviour. A whole field is just opening up, and that, therefore, we have programmes and not results — particularly in the context of the most complex organism of all, *Homo sapiens* — is hardly a cause for depression. There is much truth to the old philosophical adage that the search for the truth is to be preferred to the truth itself. Certainly, it is this search that is the chosen lot of the scientist and very exciting choice it is too.

   The chosen lot of the philosopher is something rather different, — or at least the aspect of the truth which he (or she) seeks is not that sought by the scientist, as I have been at pains to prove in previous essays. Not that this is to say that the scientist never leads off into philosophy, just as philosophers frequently, too frequently, plunge into science, as I have also been at pains to prove in previous essays! Both sides share a common belief that they can do the other's work better than the other can him (or her) self. And certainly sociobiologists are no exception in this respect. Wilson seems fairly certain that, with a few broad strokes, he can put to rest some of the major conundrums of philosophy, of ethics in particular (Wilson, 1975a, 1978). Since I am myself a philosopher, it will therefore be appropriate to conclude this essay with a brief glance at the relevance, or believed relevance, of sociobiology to ethics.

   Wilson begins *Sociobiology* as follows:

Camus said that the only serious philosophical question is suicide. That is wrong even in the strict sense intended. The biologist, who is concerned with questions of physiology and evolutionary history, realizes that self-knowledge is constrained and shaped by the emotional control centers in the hypothalamus and limbic system of the brain. These centers flood our consciousness with all the emotions — hate, love, guilt, fear, and others — that are consulted by ethical philosophers who wish to intuit the standards of good and evil. What, we are then compelled to ask, made the hypothalamus and limbic system? They evolved by natural selection. That simple biological statement must be pursued to explain ethics and ethical philosophers, if not epistemology and epistemologists, at all depths. (Wilson, 1975a, p. 3.)

But, one might ask, precisely where is that pursuit for explanation going to lead one? Wilson has, I think, two answers. One yields a variety of moral relativism. The other yields a brand of evolutionary ethics. Let us take Wilson's answers in turn.

The first answer draws attention, not only to the fact that we are what we are because of evolution — our abilities to make ethical judgements, our convictions that we are right, and so on — but that, from the viewpoint of evolution and natural selection, different people have different interests. Even in the closest-knit groups this can be so. Thus, for instance, although the interests of parent and child will often coincide, as we have seen, they may not. The parent is concerned with the welfare of all its children, even to the extent (say) of sacrificing one for the sake of the others. The child, conversely, is interested in its own welfare primarily, and only indirectly in the parent's. Wilson therefore feels justified in drawing a relativistic conclusion:

If there is any truth to this theory of innate moral pluralism, the requirements for an evolutionary approach to ethics is self-evident. It should also be clear that no single set of moral standards can be applied to all human populations, let alone all sex-age classes within each population. To impose a uniform code is therefore to create complex intractable moral dilemmas — these, of course, are the current condition of mankind. (Wilson, 1975a, p. 564.)

There are a number of reasons why Wilson's conclusion does not follow, at least as straight-forwardly as he seems to think it does, apart from the need to work out in detail the extent to which different genetic interests in a family are going to lead to different desires. For a start, from the fact that people believe in things because of evolutionary developed abilities and because it is in their evolutionary interests to do so, even though others might not believe the same, if does not follow that all collapses into a mass of relativism. At least, if it does, then the rot does not stop there. Perceptions and inferences, scientific or otherwise, come close behind, for they too depend on abilities developed through evolution and presumably serve our evolutionary interests. But it is undoubtedly true that peoples with conceptions and reasoning schemes very different from our kind of scientific methodology, have seen the world very differently from us — a world, say, populated by witches, spirits, and so on — and since they have or did manage to survive for very long times, who is to say that these beliefs and methods were not in their evolutionary interests? Hence, we must allow that these beliefs were just as valid (in the most ultimate sense that one can have) as ours. But if all becomes so relative, then even Wilson's claims about ethics become relative, and so we get caught in a familiar circular paradox.

At this point, the complaint might be made that Wilson neither intends nor is he committed to such a relativism. His point is that different people have different evolutionary interests and that we should be aware of this fact when making moral judgements. Well, apart from the fact that this is not quite what he says, two points now follow. First, the possibility of a uniform moral code no longer seems quite so remote. What we seem to have is some general maxim about the need to maximize survival or reproductive interests. This starts to look fairly universal, even though in application it might mean different things for different people. But this is no more than the fact that the universal Canadian legal code falls in different ways on different people — minors and adults, first offenders and repeaters, and so on. (In fact, I regret to say, the Canadian code is not that universal. A native woman living on a reservation loses native status if she marries a non-native. The same does not hold of a man.) Second, it starts to look at though appeal is being made to a moral principle which, although informed by evolutionary theory, does not stem from an ultimate evolutionary basis. One can believe in the truth of evolutionary theory and yet deny the premise that one ought to promote, or at least not hinder, reproductive potential. In other words, Wilson's ethics does not derive directly from sociobiology.

It might be claimed, perhaps, that at this point Wilson intends no such moral maxim. However, elsewhere, in what I have called his second answer to the problems of ethics, it does seem very much that Wilson has in mind some such maxim about the need to keep the human race going in the best possible way. He holds to a maxim, that is, that does not take individual desires and interests as ultimate. Right at the end of *Sociobiology*, Wilson talks explicitly of the need to plan for the future and this certainly seems to imply riding if necessary over particular interests for a general good. Then, in a popular article explaining sociobiology, Wilson makes very clear that he wants to go beyond particular interests, writing, in a passage which seems to me to be in flat contradiction to the above quoted passage about self-interests:

The moment has arrived to stress that there is a dangerous trap in sociobiology, one which can be avoided only by constant vigilance. The trap is the naturalistic fallacy of ethics, which uncritically concludes that what is, should be. The 'what is' in human nature is to a large extent the heritage of a Pleistocene hunter-gatherer existence. When any genetic bias is demonstrated, it cannot be used to justify a continuing practice in present and future societies. (Wilson, 1975b, p. 50.)

And, as he continues to develop this theme, it becomes very clear that Wilson's fear is some overall disaster, wiping us out or at least reducing the values of society as we now know it. Rather we are urged to put sociobiology to the task of achieving 'healthier and freer societies'.

However, whilst this is all very well — desirable in fact — it is indisputable that appeal is being made to values which lie outside of sociobiology, of evolutionary theory indeed. Continued human existence is being considered a good, particularly if it has in abundance things which we prize highly. But obviously there is no attempt, or indeed could there be attempt, to derive this ethical conclusion from the foundations of evolutionary theory. The Hardy—Weinberg law may be powerful; but it is not *that* powerful! As I pointed out in the last essay, one simply cannot derive moral values from the course of evolution — otherwise one finds oneself locked into accepting such obviously false claims as that the evolution of the smallpox virus was a good thing, and its artificial elimination by the WHO a bad thing.

In short, although I would not deny that knowledge of ourselves is essential for the making of proper moral decisions, and although I would allow that possibly sociobiology can or will contribute to that knowledge, in no way does human sociobiology yield a unique path to the true foundation of moral science. Evolutionary ethics is no more valid today that it was at the time of Herbert Spencer, its first and most notorious advocate (Spencer, 1892). Sociobiologists should stick to sociobiology — if they can give us a developed science of human behaviour, this will be no small achievement.

NOTES

[1] T. H. Huxley once quipped that Herbert Spencer's idea of a tragedy was a beautiful theory destroyed by an ugly fact. Huxley himself was given to Popper-like pronouncements on science, urging the need to "Sit down before fact as a little child, [and to] be prepared to give up every preconceived notion" (Huxley, 1900, vol. 1, p. 219). Fortunately, Huxley rarely took his own advice, as is shown, for instance, in his squabble with Richard Owen over the vertebrate theory of the skull (Ruse, 1979a, pp. 142—4).

2 Non-biologists often have trouble grasping the idea of kin selection. To them it seems almost contradictory. Let me, therefore, re-emphasize that natural selection has no place for sentimentality: parenthood *per se* is not sacred. What counts in evolution is simply holding or increasing the ratio of one's own genes in the next generation. Normally, the best strategy is to do the reproducing oneself, because even relatives do not share exactly the same genes. But an odd situation like the Hymenoptera changes matters: one leaves behind more genes through sisters than through daughters. More generally, we expect kin selection whenever we can better increase gene representation by rearing and helping non-offspring rather than offspring. Obviously, when one speaks of two organisms sharing X% of their genes, one refers only to those genes in the population which vary.

IS SCIENCE SEXIST? THE CASE OF SOCIOBIOLOGY

To many people, particularly if they are white, Anglo-Saxon, male, protestant physicists, the question "Is science sexist?" really does not make much sense. They feel that the person asking it is committing what we philosophers have learned to call a 'category mistake' (Ryle, 1949). Asking whether science is sexist is a bit like asking whether or not Tuesday is tired. It is not false, or true, that Tuesday is tired. It is simply that Tuesdays are not the sort of things that can be tired. Dogs and people get tired and (by analogy) salads. Similarly, to ask whether science is sexist is to ask the wrong kind of question of the wrong kind of thing. Norman Mailer is sexist; D. H. Lawrence is sexist; and so perhaps are the Catholic and Mormon churches. Science is not the sort of thing which could be sexist.

What would be the basis of a denial like this? Fairly obviously, the argument would run somewhat along the following lines: Science, it would be said, is an attempt to describe and explain what is going on 'out there' in the physical world. The scientist attempts to master and measure the empirical data, and to bring understanding to it. As such, his/her concern is with what *is* (or was, or will be). Now labelling something 'sexist' is rather like labelling something 'horrible'. One is going beyond the purely descriptive to the prescriptive. One is saying not just how things are, but how they ought or ought not to be. In particular, in being sexist one is treating females, not merely as different from males, but as in some ways inferior. One is saying that women are not as good as men and that, therefore, they ought not to be treated as equals. (This charge could work the other way against males, but it rarely if ever does.) Thus, D. H. Lawrence is sexist because, in his novels and stories, he deliberately portrays women as secondary, as existing for men (Millet, 1970), and the Catholic church is sexist because it refuses to allow women full personhood, particularly in the matter of becoming priests. And here, so the argument would conclude, we see the incongruity of asking whether science is sexist. Science is something which cannot possible make evaluations. Sexism is something which implies an evaluation. Hence, the two terms, science and sexism, cannot be put together.

Moreover, the argument would add, as soon as we look at concrete examples from science, we see how ludicrous such a question is. One may not particularly like the laws of nature — one can well imagine a defendent in a paternity case wishing Mendel's laws were other than they are. But to talk of particular instances of science being for better or for worse, or for or against women, is ludicrous. Newton's inverse law of gravitational attraction says nothing at all about how things ought to be, and certainly nothing about the proper status of women with respect to men. Undoubtedly some male scientists are sexist in the way they refuse women scientists proper recognition, but this prejudice can, in no way, reflect itself into their science (but see Cole and Cole, 1973).

Nevertheless, despite an initial plausibility I think this argument is naive. For all the propaganda one gets from retired Nobel Prize winners and their fellow thinkers, science is simply not just a faithful reflection of reality: a paradigm of 'objective' knowledge. Scientists have to make evaluations and decisions all of the time, and these can and do reflect cultures, interests, biases, and so forth — including here moral attitudes.

9.1. HOW SCIENCE CAN SHOW BIAS

At the most basic level one has questions about which problems one (as a scientist) is going to study (Ruse, 1979a, 1980b). It is very simplistic to think that a scientist just goes into his/her laboratory on a Monday morning and then decides to study and try to explain any old chunk of the world. Scientists get fired by 'interesting' and 'important' problems, like those most likely to win them important recognition and honours (Laudan, 1977). If anybody thinks otherwise, go away at once and read the *Double Helix* (Watson, 1968). And in their very choice of problems for study, scientists can reveal moral and other evaluative biases. Suppose, for example, some person came to a granting agency and asked for a large sum of money to study Jewish nose sizes or negro penis sizes. Suppose that, having been successful in his application but unsuccessful in the initial hunt for significant differences, he came back for more money. It surely would not be long before one started to realize that all this attention to minute or undiscovered presumed differences between Jews and blacks and others revealed a not-very-nice attitude towards Jews and blacks. And the same would hold of a journal which published all of these mind-numbing findings in great detail, even though this is certainly not to deny that there are circumstances where a journal might well publish an examination of Jewish/Gentile or black/white differences — say a journal of medical genetics talking about Tay—Sachs disease. Paying attention to something shows a choice. Surely all the attention being shown in philosophical journals today about women's topics carries, in itself, the implication that women as people are more important than has hitherto been acknowledged by philosophers. Anybody who thinks that 'interesting' is not a highly desirable quality of any potential publication has never been a referee — or been refereed! (Are you really reading this book simply because of your disinterested thirst for knowledge — any knowledge?)

Another avenue through which bias can come into science is that of the language and metaphors and examples that scientists use in their work. Let me make up a just-so story for physics, in order to show not simply how such bias could come into physics but how even physics could be tainted with sexism. Suppose physicists were to discover some new elementary particle and that this particle has two forms or states: the one form is the 'stable' (physicists' language) form, because it stays put and does not change; the other form is the 'unstable' or 'unreliable' or 'flighty' form, because it zigzags all over and, as like as not, will change into something else. Moreover, the unstable form 'dithers' or 'can't make up its mind' because frequently it flips back and forth between its normal form and another form, sometimes going permanently all of the way, and sometimes not. And now suppose finally (you guessed it!) that physicists labelled the first form the 'male' form and the second form the 'female' form. Need I say more to make my point?

Of course, it will be objected that this is just a hypothetical example and that, in fact, scientists can always change their language without any change of essential meaning, thus eliminating moral biases. Perhaps this is so, although I am not quite as sanguine about the ease of change or even of its possibility in every case, as most confidently assume — a point I tried to make in Essay 4. My point is unaffected, namely that biases can come through language, metaphors, analogies, illustrations, examples — and they do.[1] If anybody doubts this, look at some nineteenth-century anthropological works on the differences between the English and the Irish: Charles Darwin's *Descent of Man* (1871) is a good starting place. My own favourite example of science of this ilk comes from the pen of one Fleeming Jenkin (1867) who, in the course of an astute critical analysis of Darwin's views on heredity, invited the reader to imagine a white man shipwrecked among the race of black savages. Purportedly showing that Darwin could not get much permanent change, however successful a new variation might be, Jenkin lamented the fact that his white man — who would undoubtedly become king and father lots of children — would hardly turn the race more than dingy brown, if that. If this is not to show an evaluative attitude, I do not know what is.

And in other ways too, scientists can and do show their biases. For instance, scientists have metaphysical commitments according to which they organize their ideas — beliefs akin to what the Kantians call 'regulative principles' (Ruse, 1980c). The biases do not necessarily involve moral evaluations or implications, but they certainly can. For instance, the implications of Darwin's naturalistic beliefs were judged by many critics to be, not merely atheistic, but in themselves downright immoral. Claiming that the hand or the eye can be a product of natural causes is to deny design, which is to deny God, and this is sinful (Ruse, 1979a).

Also, frequently, if not almost always, scientists want to go beyond the strict confines of their evidence, speculating and making hypotheses about what the evidence implies. There is nothing wrong with this. In fact, science — certainly *good* science — would be impossible if it were not done. Constantly, science must be transcending its empirical reach, otherwise it stagnates; if indeed it can function at all. However, in going beyond the evidence there must be something influencing scientists and these influences, as often as not, are ideologies — including moral ideologies. Take, for example, the case of Malthus (1826), who argued that potential population increase will always outstrip potential food-supply increase and that, therefore, one cannot hope to save the poor through universal welfare-schemes. It certainly went beyond his evidence and undoubtedly reflected his rather conservative beliefs, particularly those about the immorality of state-run programmes for the indigent. Please note that I am not, as such, chiding Malthus, or his critics. I am simply trying to show what goes on in real science. (see Inglis (1971) for more on Malthus.)

The point has been laboured enough. Science can show biases, even moral biases. It can even show biases against women. But what must be answered positively, if this essay is to have any more than mere academic interest (surely a turn of phrase which implies something about academics!), is whether there actually are bodies of scientific endeavour and output which do show systematic prejudice against women? Even if we agree with our naive physicist that not all areas of science show bias against women (and let us agree with him to avoid lengthy argument), are there, nevertheless, some areas which do?

9.2. FREUDIAN PSYCHOANALYTIC THEORY

Even someone virtually insensitive to the currents of the past decade will know that there is one area of science which many, particularly feminists, would argue shows a gross bias against females: Freudian psychoanalytic theory. (There are those who would argue that this is not genuine theory, but I shall ignore this argument here – although, for the record, I believe it to be a specious argument. See Popper (1974), Grünbaum (1977), (1979).)

Repeatedly, Freud's "perverted point of view" (Bardwick, 1971, p. 9) has been castigated as sexist: in its language, in its metaphors, in its examples, in its metaphysics, in its hypotheses beyond the evidence, and in every other possible respect. And indeed a *prima facie* case can be made for the general claim that Freud (and most of his followers) show a systematic bias against females, considering them at best inferior to the male. Apparently, central to his position on women is his belief that their "anatomy is destiny": lacking a penis, they suffer the "anatomical tragedy" of a being who is a "castrated", "maimed", "mutilated creature". Without "the only proper genital organ", woman must come to face the "fact of her own castration [and] the consequent superiority of the male and her own inferiority", even though she "rebels against these unpleasant facts". Seemingly, woman's only hope of overcoming her loss is to have a male baby of her own. "Woman is thus granted very little validity even within her limited existence and second-rate biological equipment: were she to deliver an entire ophanage of progeny, they would only be so many dildoes" (Millett, 1970, p. 185). Indeed, even when women do do something creative in its own right, it tends to have a dubious causal origin. The female skill at weaving is intimately connected with the matted nature of her pubic hair. (The various quoted passages, with one exception to be noted, are taken from Freud's most direct writings on women: Freud (1927), (1932), (1933).)

Frankly, I tend to distrust trendy ideologies as a matter of principle. Certainly, I have a suspicion that the case against Freud is nowhere like as water-tight as the above quotes might lead one to conclude. It is amusing to note, in fact, that although "anatomy is destiny" certainly catches Freud's position on the psycho-sexual development of women, the phrase was actually used by Freud in an essay on men! Freud was pointing out the anatomical connections between the erotic and the excremental, and thus he was explaining the sexual connections.

Without really engaging in the debate, let me say two things in Freud's defence. First, one must recognize that the time and the society within which he was writing was a time and society which really did put an emphasis on the value of being male. Boys were preferred to girls, and men did things that women could not. Before condemning Freud, one needs detailed examination of his place and times. If women in Vienna at the turn of the century, regretted the lack of a penis, then Freud's pointing this out is not sexism – it really is non-tainted science (Mitchell, 1972, 1974; Strouse, 1974). Second, let us not pretend that Freud paints a picture of the male much superior to the female. Little girls may go around regretting the loss of a penis; little boys go around terrified that they will be next! (Freud, 1905). Boys are scared to death that father will cut off their penis if they copulate with their true love – mother. Indeed, some are so scared that they can never have mature heterosexual relations. In fairness to Freud, if one is to judge him on his views about women, one should judge him also on his views about men. (See also Essay 10.)

But Freud is not really my subject in this essay. Whether one decides for him or whether one decides against him, mention of his ideas and his critics does show that the topic of sexism in science is a live issue today. And it also shows the kind of form the charge might take, and even more pertinently how one might start to mount a defence. Especially, if one can establish that the women Freud was describing really did behave and feel as he claimed, then many of the harsh judgements of our day are blunted. More generally, one suspects that, if a certain science is accused of bias, say sexism, an effective defense can best be mounted, not by denying that science can ever show bias (for this is certainly not true), but by showing that the disputed claims have at least as much ground in hard empirical evidence as in the hopes and wishes of the scientist proposing them.

Keeping this fact in mind, let me now turn to my chief concern, namely that, in recent years, the charge of sexism has been brought against parts of biological science as well as parts of social science. In particular, as noted in the last essay, the new discipline of sociobiology, specifically human socio-biology, has been accused by its critics of being systematically biased against females in every possible respect (Allen *et al.*, 1975, 1976, 1977). This then is my main question: Is sociobiology sexist?

9.3. THE SOCIOBIOLOGY OF HUMAN SEXUALITY: WILSON

We know now that sociobiology is the systematic study of animal social behaviour from an evolutionary perspective. Had sociobiology stayed just at the animal level, I doubt that most of us would have heard of it – it certainly would not have got (as it did) on to the cover of *Playboy*! (Morris, 1978). But, as we saw in the last essay, sociobiologists reached up at once, causing controversy by applying their models and findings directly to our own species, *Homo sapiens*. Thus, for instance, what we find is that, having surveyed the animal world from insects up to primates, in his major work *Sociobiology: The New Synthesis*, E. O. Wilson capped off his treatment by arguing that much that had gone before continues on through and into humans. We too are motivated by our genes, as they have been selected and preserved by natural selection — individual selection that is.

And here we come to the reasons for the charge of sexism. Wilson and other sociobiologists have argued that not only are women different from men, physically and psychologically, but that many of the differences are less reflections of social conditioning and more of biology. In some sense, women are naturally the physically weaker sex, and together with this the more monogamous, the homebodies, and so forth. The genes make us this way. But as can be imagined, views like these have led critics to put sociobiology in the same category as Freudian psychoanalytic theory. It is argued that the sociobiology of sex is in no way a true reflection of reality; it is rather a figment of chauvinistic male scientists' warped minds. In order to consider this charge in fair detail, let us first look more closely at the sociobiology of human sexuality. (In what follows, I shall obviously be considering primarily the sociobiology of *hetero*sexuality. I dealt briefly in the last essay with the sociobiology of *homo*sexuality.)

In *Sociobiology: The New Synthesis* (1975), necessarily, Wilson could treat humans, including their sexuality, only very briefly. Pertinently, he claimed that we show "aggressive dominance systems, with males dominant over females" (Wilson, 1975a, p. 552), and in a little more detail he wrote as follows:

The building block of nearly all human societies is the nuclear family . . . The populace of an American industrial city, no less than a band of hunter-gatherers in the Australian desert, is organized around this unit. In both cases the family moves between regional kin by means of visits (or telephone calls and letters) and the exchange of gifts. During the day the women and children remain in the residential area while the men forage for game or its symbolic equivalent in the form of barter and money. The males cooperate in bands to hunt or deal with neighboring groups. If not actually blood relations, they tend at least to act as 'bands of brothers'. Sexual bonds are carefully contracted in observance with tribal customs and are intended to be permanent. Polygamy, either covert or explicitly sanctioned by custom, is practiced predominantly by the males. (Ibid, p. 553—4.)

Wilson has returned to the question of human sexuality in much more detail in his recent work, *On Human Nature* (1978). As mentioned, this work, devoted exclusively to humans, is very much an expansion rather than revision of the ideas Wilson was able only to sketch in *Sociobiology*. However, it does

have the virtue that one can see much more clearly why he is led to the
conclusion that he is, and this is particularly so of his discussion of sexuality.

Why would a sociobiologist like Wilson want to see biological differences
between men and women? Admittedly there are physical differences, and
admittedly both in our own society and those of pre-literate societies there
are psychological and social differences. Admittedly these psychological and
social differences do even involve males being more aggressive than females.
Males do the hunting and the fighting; females do the gathering and the bulk
of the child rearing. Males tend to be more polygamous, aggressive sexually,
and ready to shop (and sleep) around. Males do the raping; males frequent
prostitutes; and so forth. All this we can allow. But why persist, as Wilson
does, in arguing that these psycho-social differences are a function of the
genes rather than conditioning and of the environment generally? Why not
accept that men and women behave in the way that they do, entirely because
this was the way that they were taught?

The answer goes back to the fundamental sociobiological premise discussed
in the last essay: animals, including human animals, must be seen as a function
of evolution through *individual* selection, where this term obviously is being
used in opposition to group selection and thus embraces such extensions as
kin selection. We must not assume that sauce for the goose is sauce for the
gander, or rather that selection for the goose is selection for the gander. In
particular, Wilson argues that males and females (including humans) have
different 'reproductive strategies'. Confining ourselves to humans, although
we may see boys and girls, men and women, falling in love, mating, and
raising children together, we should not think that selection has designed us
so we must act as a pair or a group. Any cooperation we may see, although
it might be quite sincere, must be such that it will rebound more favourably
on us than if we did not cooperate at all. That is to say, we must be able to
raise or at least produce more offspring by cooperating than by not.

But there is more to the full story of sexuality than this. For reasons or
causes which take us back to evolutionary antiquity, the physiology of males
and females is different. Women produce only a limited number of eggs (ova),
and they can be fertilized and bear and raise children an even more limited
number of times. Moreover, once they are fertilized women are stuck with
baby, at least until they give birth — and by that time they have already put
time and effort into child-rearing. Furthermore, there is the inconvenient fact
that there is little doubt about who the mother of the child is. A mother can
hardly pretend that the baby is someone else's, and get them to take over the
onerous task of raising the child from birth to post adolescence. The only

exception would be close blood relatives who, in some circumstances, would do such raising as a function of kin selection. (I am, of course, just sketching the basic biological pattern. Sociobiologists would not deny that culture can and does come into play, perhaps masking biology sometimes, as when in some states there is care of orphans. Although in this context, apart from obvious reciprocal altruist explanations, it is worth remembering just how close many societies are to their biology: *Oliver Twist* contains more fact than fiction.)

The upshot of all the above-mentioned facts in Wilson's opinion is that the (biologically founded) psychological make-up of women is bound to be such as the aid their physiology. Women are going to be cautious about reproduction, both because it is in their interests to mate only with those men who will help produce good genetic stock and because, given the incredible amount of work involved in raising human children, they will want mates who (normally) can reasonably be expected to help with bringing up the offspring. Moreover, in some circumstances, because it is in their biological interests, women might be prepared to mate polygamously because, although they have to share a mate, at least they are going with a 'proven winner'. Better to share the top man in a society than to have the bottom man exclusively. (This is known as the 'Orians–Verner Model' (Orians 1969).) And, of course, in other respects, women's reproductive physiology will affect them physically and psychologically. If one is bound to be doing a lot of child-rearing, then adaptations for hunting and agression are hardly of first importance. Better that one be more domestic.

The story is otherwise for men. They are hardly in the same class as the boar or the stallion, but they can certainly reproduce to a virtually limitless extent. Men can go from woman to woman. Furthermore, it is not easy to pin paternity on them. (At the time of our evolution, sophisticated blood-tests were not generally available.) This means that men do not need the special adaptations of women, like sexual caution.

However, men do need special adaptations of their own. In practice, unrestrained promiscuity is not possible. On the one hand, there are other men competing for the women. On the other hand, the women themselves are looking out for their own interests and being choosy. This means that men have to settle for far less than their limit. They can have one or (at most) a few women, and they have to get involved in child-care (which is, of course, in their own interests). One special adaptation men need, therefore, is a willingness to help in child care, even if not to the same extent as women. But they need others to help them compete in the sexual jungle. They need the ability to be the sexual aggressors, on the lookout for potential fertilizees. They need the willingness to inseminate someone else's mate if they can get away with it. They need the unwillingness to have this happen to their own mates! In addition, free of the most immediate cares of child-rearing (e.g., breast feeding) and faced with the need to strive against other men, both for mates and (once successful) to succour and aid their own mates and children, men need to have evolved a biologically-based aggressiveness and dominance attitude.

In short, for Wilson the sociobiology of males and females puts them apart. Although I think it is true to say that Wilson sees the distinctive characteristics of human males and females as being things which were primarily selected during our evolution when humans lived in small bands and when the greatest threat was from the members of rival bands, he clearly thinks that the general picture holds of all societies, including our own. Nevertheless, psycho-social characteristics are most visible today and in the immediate past in preindustrial societies which are, almost by definition, those closest to nature. In summary Wilson writes:

The anatomical difference between the two kinds of sex cell is often extreme. In particular, the human egg is eighty-five thousand times larger than the human sperm. The consequences of this gametic dimorphism ramify throughout the biology and psychology of human sex. The most important immediate result is that the female places a greater investment in each of her sex cells. A woman can expect to produce only about four hundred eggs in her lifetime. Of these a maximum of about twenty can be coverted into healthy infants. The costs of bringing an infant to term and caring for it afterward are relatively enormous. In contrast, a man releases 100 million sperm with each ejaculation. Once he has achieved fertilization his purely physical commitment has ended. His genes will benefit equally with those of the female, but his investment will be far less than hers unless she can induce him to contribute to the care of the offspring. If a man were given total freedom to act, he could theoretically inseminate thousands of women in his lifetime.

The resulting conflict of interest between the sexes is a property of not only human beings but also the majority of animal species. Males are characteristically aggressive, especially toward one another and most intensely during the breeding season. In most species, assertiveness is the most profitable male strategy. During the full period of time it takes to bring a fetus to term, from the fertilization of the egg to the birth of the infant, one male can fertilize may females but a female can be fertilized by only one male. Thus if males are able to court one female after another, some will be big winners and others will be absolute losers, while virtually all healthy females will succeed in being fertilized. It pays males to be aggressive, hasty, fickle, and undiscriminating. In theory it is more profitable for females to be coy, to hold back until they can identify males with the best genes. In species that rear young, it is also important for the females to select males who are more likely to stay with them after insemination.

Human beings obey this biological principle faithfully. (Wilson, 1978, pp. 124—5.)

9.4. THE SOCIOBIOLOGY OF HUMAN SEXUALITY: SYMONS

Wilson is not the only sociobiologist, although it sometimes is difficult to realize this from the popular press. Other sociobiologists have looked at human sexuality, and there has just appeared a full-length work on the topic by the primate biologist, Donald Symons (1979). As might be expected, although there are some differences, Symons' theoretical and empirical stance is very similar to Wilson's. What he does try to do is work out many of the basic ideas in much more detail. It is neither possible nor necessary to cover everything discussed by Symons. But let me mention three items to give a general flavour, and to round out my presentation of the sociobiology of human sexuality.

First there is the matter of *sexual arousal*. Although some recent psychological studies have been interpreted otherwise, everyday and scientific evidence implies that men are much more readily aroused sexually than women (Kinsey *et al.*, 1948, 1953). The pornography industry exists almost entirely for men — indeed some magazines for women which used to feature nude males no longer bother to do so (U.S. Commission on Obscenity and Pornography, 1970). Even those, which still do feature nude males, find their primary market amongst male homosexuals. Moreover: "The most striking feature of pornotopia is that sex is sheer lust and physical gratification, devoid of more tender feelings and encumbering relationships, in which women are always aroused, or at least easily arousable, and ultimately are always willing. There is no evidence that a similar female fantasy world exists ..." (Symons, 1979, p. 171).

Now many researchers, particularly social scientists, argue that these differences are all laid on by culture concealing a deeper reality in which males and females differ little, if at all. (I take it that this would be essentially the position of social-learning theorists and of cognitive-developmental theorists, for instance Bandura (1969) and Kohlberg (1969), respectively. It is also the position of feminist psychologists, for instance Frieze *et al.* (1978) and Brooks-Gunn and Matthews (1979).) However, Symons argues that the differences are much more plausibly interpreted as part of our biology, particularly given the fact that the anthropological evidence is that it is males who are erotically excited by females, rather than vice-versa. Many preliterate societies openly allow males to parade around with their genitals in full view; very few societies allow women to have uncovered genitals. Those which do permit full naked females have rules or customs prohibiting males from staring or females from openly displaying their vaginas (e.g., as when sitting).

Symons writes:

Male-female differences in tendencies to be sexually aroused by the visual stimulus of a
member of the opposite sex – whether this stimulus is a drawing, painting, photography,
or actual person – can be parsimoniously explained in terms of ultimate causation, although their proximate bases remain obscure. Because a male can potentially impregnate
a female at almost no cost to himself in terms of time and energy, selection favored the
basic male tendency to become sexually aroused being proportionate to perceived female
reproductive value; for a male, any random mating may pay off reproductively. . . .
Human females, on the other hand, invest a substantial amount of energy and incur
serious risks by becoming pregnant, hence the circumstances of impregnation are extremely important to female reproductive success. A nubile female virtually never
experiences difficulty in finding willing sexual partners, and in a natural habitat nubile
females are probably always married. The basic female 'strategy' is to obtain the best
possible husband, to be fertilized by the fittest available male (always, of course, taking
risk into account), and to maximize the returns on sexual favors bestowed: to be sexually
aroused by the sight of the males would promote random matings, thus undermining all
of these aims, and would also waste time and energy that could be spent in economically
significant activities and in nurturing children. A female's reproductive success would be
seriously compromised by the propensity to be sexually aroused by the sight of males.
(Symons, 1979, p. 180.)

Next there is the question of *sexual choice*. Here Symons is very much in
line with the thinking of other sociobiologists, including Wilson. Biologically
speaking, there is no reason to think that male and female strategies, with
respect to the ideal marriage or sex partner, will be the same. Indeed, biology
suggests the kinds of differences which are reported by anthropology and
which, in fact, hold in respects in our own society. For men, the ideal partner
is one which, if fertilized, will have the best chance of conceiving and rearing
fit children. This means that there is a premium on relatively young, healthy
women – and this, in fact, is that we find attracts men. Actually Symons
suggests that the ideal marriage partner and the ideal sex partner might not be
the same. A woman in her mid-twenties is best for random mating, because
she has optimum chance of raising children. A woman somewhat younger is
better for a wife, "since a male who marries a female of this age maximizes
his chances of tying up her entire reproductive output" (ibid., p. 189). Moreover, what really turns men on is a nice, new partner. This is in line with their
biology, because all new women are potential fertilizees.

Conversely for a woman, what is important is not the ability to conceive
healthy children – men do not conceive, and in this respect one man's sperm
is much like another man's – but is the ability to aid her and her children.
This can be done both through the conferring of admirable hereditary traits
and aid in child-rearing, which traits and aid could be direct or indirect. All of

this adds up to the fact that age is not the crucial factor for the choice of men, but status is. Moreover, "from the female's point of view, a high-status male is both the best choice for a husband and for a sex partner" (ibid., p. 193). And once having got herself a winner, a woman's best bet usually is to stay with him. Adultery for her pays only if she can upgrade the genes her offspring receive, or if (on a more permanent basis) she can get better child-care or help.

Finally, let us mention Symons' discussion of *copulation as a female service*. Both males and females enjoy sex. If anything, women have the capacity to enjoy it more than men, because they can have multiple orgasms in the same bout of sexual intercourse (Masters and Johnson, 1966). And yet it is virtually always men who have to beg and ask for sex, and to give bribes or presents. It is virtually never women. Prostitutes exist almost exclusively for men. The exception proves the rule: gigolos are for old women who have lost their sexual attractiveness.

Why this uni-directional relationship should obtain has often puzzled anthropologists. Listen to Malinowski on the Trobriana islanders:

In the course of every love affair the man has constantly to give small presents to the woman. To the natives the need of one-sided payment is self-evident. This custom implies that sexual intercourse, even where there is mutual attachment, is a service rendered by the female to the male . . . – This rule is by no means logical or self-evident. Considering the great freedom of women and their equality with men in all matters, especially that of sex, considering also that the natives fully realize that women are as inclined to intercourse as men, one would expect the sexual relation to be regarded as an exchange of services itself reciprocal. But custom, arbitrary and inconsequent here as elsewhere, decrees that it is a service from women to men, and men have to pay. (Symons, 1979, p. 254, quoting Malinowski, 1929.)

Symons' argument is that there is nothing at all arbitrary and inconsequential here. The one-way flow of gifts and requests stems from biology.

Copulation as a female service is easily explained in terms of ultimate causation: since the minimum male parental investment is almost zero, males stand to benefit from copulating with any fertile female (if the risk is low enough), whereas females do not stand to benefit reproductively from copulating with many males no matter what the risk is. (Symons, 1979, p. 261.)

In other words, man's begging or giving payment for sex and woman's being begged or accepting payment for sex is a simple consequence of the unequal biological costs of sex. It costs a man virtually nothing; it can cost a woman a great deal. It is, therefore, in a woman's biological interests to be cautious and to expect some returns to help redress the balance. Payment is not simply a question of a crude dominant male corrupting an innocent woman with filthy lucre or its equivalent.

9.5. IS SOCIOBIOLOGY SEXIST? THE LESSER CHARGES

Enough has been said to make clear the sociobiological view of sex: the critics will claim that more than enough has been said. Let me turn now to the charges of sexism. First, in order to clear the air, let me deal with the lesser charges. (Speaking of 'lesser' and 'major' is my ranking.) These are charges that the critics brought against *Sociobiology*, and which, to be honest, certainly do not strike me as marking sociobiology as a subject as irrevocably sexist. Indeed, in some respects they seem to me to be a little unfair.

One objection that the critics (Allen *et al.*, 1976, 1977) had was that throughout *Sociobiology* Wilson referred to humans of both sexes as 'he' and 'his', and as 'men'. This they felt was to give a not-too-subtle slight to women. However, although it is indeed true that Wilson did do this, one hardly feels that this convicts him, or more importantly, convicts sociobiology, of absolute and unremovable sexism. Even today matters are still in flux, and many would continue to argue that referring to all humans as 'men' has no implications at all about the status of women. In any case, if one wants to eliminate these supposedly offensive terms, one can surely do so without in any way affecting the content of sociobiology in general and of Wilson's presentation in particular. Incidentally, people who live in glass houses should not throw stones. One of the most vociferous critics of sociobiology, Richard Lewontin, had himself, until the year before the appearance of Wilson's book, used exactly the same, supposedly sexist language (Lewontin, 1974).

About on a par with this criticism about the pronouns was the objection that Wilson showed bias against females because in all of his pictures (in *Sociobiology*) males are put prominently in the foreground. In fact, this is not true and even if it were, hardly shows sociobiology itself to be irredeemably sexist. Incidentally, the pictures were drawn by a woman. No doubt the critics would claim that this confirms the sad state at which affairs have arrived (Figure 9.1).

A somewhat more serious criticism is that, in his metaphors and his illustrative examples, Wilson shows his prejudice against women. For instance, great play is made of the fact that Wilson likens courtship displays to contests involving 'salesmanship' and 'sales resistance', with the courted sex developing 'coyness'. This is thought to cast women in a rather degrading light. However, it is worth quoting in full the pertinent, supposedly offensive passage. When

Fig. 9.1.   One of the supposedly sexist pictures in Wilson's *Sociobiology* (pp. 506–7), drawn by Sarah Landry.

this is done, one can see clearly that things are not quite so straightforward. If anything, Wilson is being rather misrepresented.

Pure epigamic display can be envisioned as a contest between salesmanship and sales resistance. The sex that courts, ordinarily the male, plans to invest less reproductive effort in the offspring. What it offers to the female is chiefly evidence that it is fully normal and physiologically fit. But this warranty consists of only a brief performance, so that strong selective pressures exist for less fit individuals to present a false image. The courted sex, usually the female, will therefore find it strongly advantageous to distinguish the really fit from the pretended fit. Consequently, there will be a strong tendency for the courted sex to develop coyness. That is, its responses will be hesitant and cautious in a way that evokes still more displays and makes correct discrimination easier. (Wilson, 1975a, p. 320.)

Three points emerge. First, it is clear that Wilson does not always see the female as the 'coy' sex. It all depends on who is going to make the most reproductive effort. Admittedly, it is usually the female who makes the most effort, and as another passage quoted above makes clear, Wilson does think it is the human females who are the primary workers and thus more 'coy' (Wilson, 1978, p. 125); but things can be reversed. In the fish, in fact, things are changed round and it is the male who is 'coy' and who is courted (Dawkins 1976). Hence, Wilson is certainly not picking out human females for special degrading treatment.

Second, obviously Wilson is using metaphors. There is no real selling going on; no money changes hands. But why should he not use these metaphors?

The male is trying to get a message across quickly and if necessary cover up his deficiencies. Shades of automobile dealers showrooms! The female has got to be cautious and discriminate between the genuine article and phonies. What is this but 'coyness' in a sense. Of course, Wilson may be wrong in implying that this is part of the biological programming of the human female — more on this later — but describing behaviour in such a way is not sexist.

Third, as in the Freudian case, males are hardly being portrayed in that favourable a light. They are lazy, letting females do the work, trying to cover up their deficiencies. Significantly, Wilson (1975a, p. 504) refers to male lions as 'parasites'. In short, they are hardly the paradigm of something to be admired, valued, and emulated.

But in a sense, this is all preliminary skirmishing. Suppose that we get rid of all of the offensive language and examples and metaphors (whether they are really offensive or not). Although it may not always be possible, in the present case, this is something we can surely do without drastic change in meaning. In fact, Wilson's *On Human Nature* and Symons' *The Evolution of Human Sexuality* are much more careful in this regard. There are, for instance, no pictures in either! (although see Figure 9.2). But what now of sociobiology? Is it still sexist? There is little doubt that the critics would (and do) argue that it is. Let us, therefore, examine what we might call the major or more fundamental charge.

9.6. IS SOCIOBIOLOGY SEXIST? THE MAJOR CHARGE

The major charge can be put as follows: All the talk about language and metaphors obscures the truly pernicious nature of human sociobiology. What we must recognize is that, both in theory and in the selective collection of empirical facts, sociobiology portrays women (as opposed to men) in a lesser light. They are the domestic homebodies, condemned by their biology always to be so. Males are the aggressors, both in fact and by right. Moreover, the implication is that what is biologically 'natural' is also morally 'natural'. (We have an ambiguous play akin to that which occurs in the term 'natural foods' — what a wholesome image this conjures up!) The Germans were right when they consigned women to the superstitious drudgery of life: *Kinder, Kirche, Kuche.*

But, the critics will argue, this picture of men and women is something justified neither by theory nor by fact. It is simply a blatantly ideological wolf dressed up in scientific lamb's clothing. As one critic of Symons, the social scientist Clifford Geertz, has already said: "The moral equivalent of

Fig. 9.2. The cover of Symon's *The Evolution of Human Sexuality*. The picture is of a (single) human ovum surrounded by many, many, sperm; however one must in fairness note that the ovum is eighty-five thousand times larger than the sperm, and one doubts there are that many sperm in the picture.

fast-food, Symons' book is not artlessly neutral, it is skillfully improverished" (Geertz, 1980, p. 4). If particle physics is the *filet mignon* of the scientific world, then human sociobiology is the Big Mac.

One thing is clear. One possible line of defense just will not do. One cannot argue that, appearances to the contrary, the sociobiologists really consider males and females the same. The whole point of the sociobiology of sex is that they do not. Males are males and females are females, and the wonder is that occasionally the twain do meet. It seems, therefore, that the only real line of defense against the charge of sexism is to show that the sociobiology of sex has some fair claim to being plausible. If women really do have an anxiety about having no penis and if this is a significant factor in their psychosocial makeup and behaviour, then one can hardly accuse Freud of being a sexist in drawing attention to the point: at least not in the context of an overall analysis of human sexuality. Analogously, if men and women really do respond to pornography in different ways because of their genes, then one can hardly accuse Symons of being a sexist in drawing attention to this point: at least, not in the context of an overall analysis of human sexuality. I take it incidentally, that no one could seriously argue that sociobiologists show bias in even wanting to consider male/female differences. As mentioned in the case of homosexuality in the last chapter, sociobiologists are Darwinian evolutionists and, hence, for them the name of the game is reproduction. They would be remiss were they not to try to explain the facts of sexuality.

I think one must be fair in one's assessment of evidence and plausibility. One must not set unreasonably high standards for sociobiology. One cannot expect sociobiology (i.e., human sociobiology) to be proven absolutely beyond the shadow of a possible doubt. To set a criterion like this would be to exclude most science and, certainly, all social science. What one must be able to show is that, as things stand at the moment, human sociobiology is a reasonable hypothesis. It may be one hypothesis among others, but if it can hold its own at this level then I suggest that this is good grounds for denying the charge of sexism. One may not much like the situation and think that this all goes to show that God was a bit of a sexist — but then, we all know what sex He was.

Is human sociobiology, more specifically the human sociobiology of sex, a reasonable hypothesis? I shall assume without argument that the empirical facts are very much as been presupposed, namely that in both preliterate and industrial societies (although to different degrees) men and women tend to differ psycho-socially. The occasional counter-example is not that bother-some; at least, no more so than the counter-examples of the physical sciences. Men are more aggressive, the initiators in sex, those with the roving eyes, the users of prostitutes, and so forth, as compared to women. Even the critics seem to allow this much. " . . . it must be recognized that sexism, the *socially prescribed* power men have over women, does exist in most societies" (Allen *et al.*, 1977). The question is whether the genes play a crucial causal role in the differences, or whether they are all a result of social prescription.

I am afraid that one has to face up to and accept one obvious point, namely that none of the sociobiologists have much idea about which genes might be at work in the sociobiology of sex; although, as we shall see, this does not necessarily imply entire ignorance of proximate causes. The evidence must, therefore, all be somewhat indirect. But this is hardly cause for absolute rejection or despair. Direct evidence is really not that commonly available in science. Physicists do not watch sub-atomic particles themselves dashing around, nor do geologists sit on continents and allow themselves to be carried around the globe.

I believe there are three possible sources which might incline one to take the sociobiology of human sex seriously. First, there is the fact that the sociobiology of human sex does not stand on its own: it is part of general sociobiology, specifically the sociobiology of animal sex. And sociobiology does not stand on its own: as we saw in the last essay, it is part of evolutionary theory. Now, evolutionary theory is a well-confirmed theory. (See Essay 2.) It is reasonable to assume that, in general, organic characteristics are a function of genes which have been selected for their ability to confer reproductive fitness on their possessors. Moreover, that part of evolutionary theory which is to do with the social behaviour of animals, sociobiology in general, is one which it is becoming increasingly plausible to take seriously. (See Essay 8 and Ruse, 1979b.) I do not claim that it is all well-confirmed or that all controversies are gone — they are not — but I do claim that there is increasing evidence that animal social behaviour is under the control of the genes, and that the kinds of models sociobiologists propose are appropriate tools for explanation. In particular, in the case of sex it does make sense to consider the two sexes separately and to think that different reproductive strategies follow from the different contributions of males and females. In the mammals particularly, females are bound to take the more cautious approach in sexual encounters because they are the ones who take the brunt of the child-rearing: indeed, in most animals, females do all the child-rearing. (Wilson (1975a) demonstrates this fact most effectively.)

Do not misunderstand me. I am not trying surreptitiously to slip over the sociobiology of human sexuality. The whole point about human beings is that we have culture, which allows us to transmit ideas and attitudes at a level above the genes, as it were. What I am saying is that the animal sociobiology of sex works (i.e., has plausible theories). What I would also point out is that human beings have obviously been subject to sexually dimorphic selective forces at the physical level — woman's broader hips than man's have an adaptive value — and that the physiology of humans (for instance, the difference in the sizes between the male and female reproductive contributions) is such that sociobiological factors could work. And, as pointed out above, human social behaviour is what sociobiology would lead one to expect. These factors do not make the human sociobiology of sex true. But they do start to lift it above the level of the ridiculous, particularly given that much social conditioning (e.g., Judeo—Christian religious moralizing) is designed to take us from what the biology predicts — and yet many are unmoved. Even in Puritan New England adultery was not unknown — and every other repressive society had had its Hester Prynnes. I am not claiming that culture has no effect — obviously it does. The question is whether it has total effect. Selection can work even if a gene only manifests itself occasionally.

The second possible area of support for the sociobiology of human sexuality lies in direct analogies with other animals, particularly the higher primates. It is clear that Wilson thinks there is some support here. He writes:

Characters are considered conservative if they remain constant at the level of the taxonomic family or throughout the order Primates, and they are the ones most likely to have persisted in relatively unaltered form into the evolution of *Homo*. These conservative traits include aggressive dominance systems, with males generally dominant over females, scaling in the intensity of responses, especially during aggressive interactions; intensive and prolonged maternal care, with a pronounced degree of socialization in the young; and matrilineal social organization. This classification of behavioural traits offers an appropriate basis for hypothesis formation. It allows a qualitative assessment of the probabilities that various behavioural traits have persisted into modern *Homo sapiens*. (Wilson, 1975a, p. 551.)

On the other hand, Wilson himself admits that one really cannot get too much from the primates. Among them one finds monogamy, polygamy, pair bonding, no pair bonding, male involvement in parental care, no male involvement in parental care, and most of the other options which spring to mind. Obviously, any analogies one can draw are more of heuristic value than justificatory. Significantly, Symons, who is (as mentioned) a primate biologist, is very dubious about the value of analogies from the apes to humans. "Talk of why (or whether) humans pair-bond like gibbons strikes me as belonging to the same realm of discourse as talk of why the sea is boiling hot and whether pigs have wings" (Symons, 1979, p. 108). Certainly Symons does not deny primate analogies entirely — he uses the chimpanzee as a model for our ancestors — but it is probably best to agree with him that we should not really look to the great apes for much support for the sociobiology of human sexuality.

Third and finally, we have the question of implications and predictions which follow from or can be made on the basis of the sociobiology of human sexuality. Copernicus' theory was much more plausible after Galileo predicted and discovered the phases of Venus (Kuhn 1957); Darwin's theory similarly gained credibility after Bates's work on mimicry (Ruse, 1979a). Some philosophers, for example Whewell (1840), have argued that this is the truly essential type of evidence necessary for strong theory confirmation.

Can sociobiology yield analogous predictions about human sexuality, particularly those predictions and implications which are perhaps a little unexpected and hard to explain, especially if one subscribes exclusively to an environmentalist viewpoint? The sociobiologists claim that it can. There are two major pertinent predictions which have been claimed as following from sociobiology, but not from environmental theses.

The first prediction stems from the facts that one's psycho-social attitudes and behaviour are obviously linked to the nature of the brain, and that equally obviously if sociobiology is correct then the genes of a man are going to cause a 'masculine' brain and the genes of a woman are going to cause a 'feminine' brain. If sociobiology is right, then no matter what the naturally-occurring environment, a masculine brain is going to cause male behaviour and attitudes and a feminine brain is going to cause female behaviour and attitudes. An environmental thesis implies that, no matter what the genes, if one is brought up as a male, one will behave as a male (i.e., as a male as we now understand males), and if one is brought up as a female one will behave as a female.

Now suppose nature (with or without human help) plays a trick. Suppose that a person has a brain exposed to the kinds of influences one would expect from genes of one sex, but is brought up as a person belonging to the other sex. The environmentalist would expect the upbringing to be the decisive factor; the sociobiologist the influences on the brain. In fact, there are some people who do actually fall into the kind of class being supposed here (see Money and Ehrhardt, 1972; Money and Musaph, 1977). It appears that the genes of a male cause the fetal brain to be exposed to a higher level of male sex hormones, testosterone, than do the genes of a female (specifically between the third and six months of life, when the hypothalamus is developing). However, naturally and artificially some females are exposed to the high testosterone levels during fetal growth. This means that they have 'masculinized' brains. And the sociobiologists claim that, just as they predicted (and as the environmentalists would not have predicted), such genetic females grow up with male-type behaviour — even though physically they may be female and have always been treated as such.

Thus Wilson concludes that "at birth the twig is bent a little bit" (Wilson, 1978, p. 132). One must add that the bending is not that extreme. Although experiments on rats and monkeys support the above human findings (Goy and Phoenix, 1972; Goy et al., 1977), the girls in question certainly do not slip all the way into a male role or 'gender identity'. Despite their exhibiting tomboyish behaviour and, without treatment, showing male eroticism patterns (e.g., in the nature of their sex dreams), and even being concerned far more with such things as careers than families, they do, nevertheless, have no doubt that they are female rather than male — nor do they wish to change (Money and Ehrhardt, 1972). The nature of their sexual orientation seems to be a matter of some controversy. It was believed that the girls showed no particular propensity towards lesbianism: a belief in line with the expectations of the major researchers on the effects of hormones on gender identity. Now the evidence is starting to point the other way a little, although there is certainly no absolute cause and effection relationship (Money and Schwartz, 1978). There will be a little more on this point in the next essay. For a rather more negative assessment of the effects of sex hormones on human behaviour and attitudes than that made by the sociobiologists (or that I myself would make for that matter), see Adkins (1980).

The second prediction concerns homosexuals. In the view of the sociobiologists (Symons in particular), if the environmentalist thesis is true, one might expect male and female homosexual patterns to converge. Male homosexuals, having foresworn the ultimate 'masculine' objective, namely trying to mate with women, will have no reason to be locked into other common male patterns, like looking continually for fresh sex partners.[2] Lesbians, on the other hand, will have no reason to be as reticent as their heterosexual sisters, and will go more towards the heterosexual male pattern. But, in fact, we find that what obtains in reality is what sociobiology leads one to expect. Male homosexuals, unfettered by the reticence of a female partner, are incredibly promiscuous. Lesbians are anything but.

The existence of large numbers of exclusive homosexuals in contemporary Western societies attests to the importance of social experience in determining the objects that humans sexually desire; but the fact that homosexual men behave in many ways like heterosexual men, only more so, and lesbians behave like heterosexual women, only more so, indicates that some other aspects of human sexuality are not so plastic. (Symons, 1979, pp. 304–5.)

My own feeling is that the sociobiologists can rightly claim that these two implications help to make their position more plausible than otherwise. How one critic, Geertz, can claim that Symons' views on homosexuals are "at about the level of descriptions of the Irish as garrulous and the Sherpas as loyal" (Geertz, 1980, p. 4), is quite beyond me. In fact, since Symons completed his work there has appeared the most comprehensive survey ever done on homosexuality, that sponsored by the Kinsey Institute, and it confirms absolutely the difference between male and female homosexuals. Males have vast numbers of virtually anonymous sex partners; lesbians are much closer to the heterosexual norm (Bell and Weinberg, 1978).

I would caution, however, that the implications cannot be taken as definitive. One can always defend a hypothesis by adding supplementary clauses, and I am sure that the environmentalists can and will try to show how, for instance, the data on homosexuals is consistent with their position. Moreover, I am not entirely sure how happily the two implications ride together. If there is indeed any connection between the hormones and sexual orientation, then perhaps one might expect lesbians to push more towards male patterns generally, and perhaps the converse for male homosexuals. But this is all still very much a matter of debate and research, so I certainly do not think that at this moment the two implications can definitely be said to be in conflict.

All in all therefore, in concluding this brief discussion of the evidence for the sociobiology of human sexuality, I would argue that a case has been made for its plausibility. I would not claim that it is an absolutely true or well-confirmed theory — anything but. But as I pointed out earlier, such a claim is not necessary to defend sociobiology against a charge of sexism. What is necessary is that, judged as science, the theory and its hypotheses be plausible. And this I think they can fairly be said to be.

9.7. CONCLUDING REFLECTIONS FOR THE FEMINIST

My general conclusion, therefore, is that the sociobiology of human sexuality is not sexist — it is not simply a refurbishing of an ideology demeaning to women. Against the most fundamental charge of sexism, the conclusive refutation can be made that the theory may, in fact, be true. But as I come to the end of my discussion, let me add two final points, not so much to qualify what has been said, but more to put things in perspective and to help those ardently in favour of the equality of the sexes look more favourably on sociobiology.[3]

First, note that there is nothing in sociobiology to suggest that women are inferior to men in the most desirable human characteristics — intelligence, sensitivity, artistic ability, loyalty, and so forth. There are no crude nineteenth-century claims that because a woman's brain is smaller than a man's, she is therefore not as clever (Mosedale, 1978). Indeed, there seems to me to be no implication in sociobiology that a woman would need any less intelligence than a man. Admittedly, it is true that at one point Wilson (1975b) notes empirical evidence suggesting that, whereas men have greater mathematical ability than women, women have more verbal ability than men. He then goes on to suggest that these differences may lead to men playing a more dominant social role than women. However, I confess I am not at all sure why this conclusion should be thought to follow, or indeed how sociobiology could imply the empirical findings — if indeed they are true.

Remember also what was noted above, namely that if one persists in believing that sociobiology paints a gloomy picture of the female fate, like Freudianism it really does little more for men either. They are portrayed as in a state of perpetual adolescence, always ready to hop into someone else's bed if the opportunity arises. (This is an exaggeration, but the implication is that there is a tendency this way.) And although the sociobiology of human sexuality sees male humans as getting more involved in child care than most mammalian males, their contribution to this hard work is hardly of the order of the females. The males are off fighting or competing somewhere. Biologically speaking, males are not moral paradigms. Surely this fact totally refutes the claim that sociobiologists are 'doing a job' on females simply by virtue of the fact that they pick out human sexuality for explanation.

The second and perhaps more important concluding point is that the sociobiologists openly concede the power of culture of overcome biology. As things are today in our own society, culture obviously influences our beliefs and actions, including those which are sexual. Furthermore, if one decides that a society of androgenous sameness is morally desirable, then it might be possible to attain this.

Throughout the Pleistocene boys and girls must have been reared differently, but it is most unlikely that these different rearing conditions were the sole developmental mechanism responsible for sex differences in sexuality. The data presented . . . suggest some developmental fixity in sexuality (there are no myths of sleeping males who require a princess' kiss to awaken them: if males are still dozing at puberty they are awakened by nature). Very likely, many of these sex differences would prove to be innate . . . if the environment were held constant. That is, males and females exposed to identical environmental conditions during ontogeny would develop different sexual behaviours, attitudes, and feelings. This does not necessarily mean that it would be impossible to rear boys and girls so that they developed identical sexualities, but simply that identical sexualities would not result from identical rearing conditions. (Symons, 1979, p. 307.)

In other words, by paying deliberate attention, we might be able to overcome our biology. The sociobiology of human sexuality does not, therefore, extol what is (or what is suggested to be) as good, or as inevitable. It is not a doctrine for keeping people in their present state. What it does suggest, rather, is that we are never going to get people out of their present state unless we realize the causes for the state. I suppose it does suggest that change might not be as straightforward was many environmentalists suggest or without possible deleterious side-effects. But then, things rarely are as straightforward as environmentalists suggest. The important point is that sociobiology, generally and as it applied to human sexuality, does not imply that we are necessarily and forever prisoners of our Pleistocene past.

Indeed, it seems worth making the point that with very little difficulty, the position of the sociobiologists can be made to mesh with the most modern thinking by sexologists on the development of gender identity — the feelings and inclinations which are part of the sense of being either male or female. The position of Robert Stoller (1968), for instance, is that we have a certain biological disposition towards a gender identity (XY babies towards maleness and XX babies towards femaleness), and that then this is reinforced (or opposed) by environmental factors. It might perhaps be that the sociobiologists give biology a bigger role than do the sexologists, but I would point out that there is nothing in sociobiology which denies the importance of environmental factors or that they can aid adaptive value. In short, if one could suggest that the environmental factors reinforcing gender identity are very stable, then from a sociobiological viewpoint, they are just as adequate as direct wiring of the brain by the genes (e.g., through hormones). But, in fact, the position of the sexologist seems to be that the factors are stable; although admittedly they do break down sometimes, as of course can happen also with hormones. The chief factors supposed are one's own reactions to one's own genitals, and parental behaviour towards the very young child — and both of these factors are indeed very stable (Stoller, 1968, 1976; Green, 1974). Hence, if one believes that the differences in gender identities (and the ways that these manifest themselves in gender roles) are biologically adaptive, I think one can be both a sociobiologist and accept almost all the recent thought by more conventional sexologists about the development of the sense of maleness and femaleness.

Incidentally, this conclusion, that successful human sociobiology supports rather than excludes other explanatory approaches to human behaviour, is very much in line with themes I have developed more extensively elsewhere (especially Ruse (1979b)). If human sociobiology works, although we will certainly have replacement of extreme environmentalist claims, and although we may have some reduction, the sexuality case suggests that, generally, what we might look for is a melding of the biological and the social. Biologists emphasize always that biological characteristics, including behaviour, are a product of the genes *together with* the environment. In other words, although one can understand the fear that social scientists have of sociobiology, the final synthesis will quite possibly be one where biology and social science will be equal and essential partners. I might add as one final word however that although sexologists believe that environmental factors can greatly alter the development of gender identities, they believe also that if this happens, people tend to be very unhappy. Hence, if one believes that human happiness is a desirable phenomenon, far more than the sociobiologists the sexologists seem to warn of the dangers of wholesale attempts to eliminate the differences between men and women. I am glad to say that discussion of the desirability of sweeping programmes of social change, including those aimed at total human androgeny, really is beyond the scope of this essay.

NOTES

[1] Having made up my little just-so story, I was delighted to find that a recent work on solid-state physics is about 'ill-condensed matter'! This refers to 'disordered materials', about which one reviewer writes that: "The pejorative flavor of 'ill' is not entirely inappropriate, disorder has long been viewed as slightly unclean and certainly has been a source of great difficulty in both experimental characterization and theoretical understanding" (Palmer (1980) referring to Balian *et al.*, (1979)).

[2] It is, of course, a loaded question today whether (male) homosexuality is unmasculine. Let us side-step controversy by simply acknowledging that male homosexuals are not attracted by the same sex objects as the average male.

[3] This may be a vain hope. Elsewhere, in the course of a defense of sociobiology, I suggested that generally women might make better doctors than men (Ruse, 1979b). For my pains, I was labelled a sexist (Darden, 1980). Had I suggested that women might make better nurses than men, or better air hostesses, I could understand the reaction. But doctors?

ARE HOMOSEXUALS SICK?

There is much controversy today about whether a homosexual orientation is in itself a disease or sickness. If one is attracted sexually to members of one's own sex rather than to members of the opposite sex, then is this a sign that one is ill — standing in need of a cure? Or is it the case that having a homosexual orientation is simply a matter of having an attribute different from heterosexuals — something on a par with having blue eyes rather than brown?

Not surprisingly, militant homosexuals tend to see homosexuality as nothing more than a variant form of sexual orientation. "I have come to an unshakable conclusion: the illness theory of homosexuality is a pack of lies, concocted out of myths of a patriarchal society for a political purpose. Psychiatry dedicated to making sick people well has been the corner-stone of a system of oppression that makes gay people sick" (Gold, 1973, p. 1211). But there are medical people who argue in much the same way. The influential psychiatrist Judd Marmor states: "Surely the time has come for psychiatry to give up the archaic practice of classifying the millions of men and women who accept or prefer homosexual object choices as being, by virtue of that fact alone, mentally ill. The fact that their alternative life-style happens to be out of favor with current cultural conventions must not be a basis in itself for a diagnosis of psychopathology. It is our task as psychiatrists to be healers of the distressed, not watchdogs of our social mores" (Marmor, 1973, p. 1209).

Conversely however, Irving Bieber states that "homosexuality is not an adaptation of choice: it is brought about by fears that inhibit satisfactory heterosexual functioning" (Bieber, 1973, p. 1210). Similarly, Charles Socarides states that "homosexuality represents a disorder of sexual development and does not fall within the range of normal sexual behaviour" (Socarides, 1973, p. 1212). And in line with the view of these two psychiatrists, the endocrinologist Gunther Dörner speaks of homosexuality as involving "inborn disturbances of gonadal functions and sexual behaviour in man", and he thinks it is a "sexual deviation" in need of cure (Dörner, 1976). As is well known, these differences of opinion are wide-spread. On 15 December 1973, the trustees of the American Psychiatric Association decided to stop listing homosexual orientation as a mental disorder, a decision ratified shortly after

by the Association as a whole by a vote of 5854 to 3810. Undoubtedly this
vote represents changing opinion which is nevertheless still very divided.
(See the *New York Times*, 16 December 1973 and 9 April 1974. For more
conclusions on the health/sickness status of homosexuality, with varying
degrees of supporting argument, see Green (1972), (1973), Stoller (1973),
Spitzer (1973).)

Obviously, whether or not homosexuality is to be judged a sickness or a
disease or an illness depends in part on the facts, both at what one might call
the 'empirical' or 'phenomenal' level and at the causal level.[1] One wants to
know something about homosexuality, how it affects people, and what
putative causes have been proposed for it. But there is more than this. One
must also learn how terms like 'disease' and 'illness' are used. Only when one
has done a philosophical analysis at this level can one then examine the
empirical and causal claims and make a proper judgement about how to
assess the status of people's sexual orientations and, in particular, to make a
judgement about the healthiness or sickness of homosexuality.

This then is my task in this, the final essay of this collection. I aim for a
philosophical-cum-scientific analysis of the health/sickness status of homosexuality. Because this is somewhat of a preliminary inquiry, I cannot pretend
to offer definitive answers of my own, but I do want to learn why people give
different answers and what sorts of assumptions would be directing them to
these different answers. In a sense, I want to help the reader make up his/her
own mind, although I shall certainly not make a rigid attempt to conceal my
own feelings and suspicions entirely. I shall begin with brief exposition of two
recently proposed philosophical models of health, disease, and illness. I shall
not be directly critical of these models, although I do think that, in the
course of the discussion, certain strains will appear in at least one of the
models. Then I shall run quickly both through the three main categories of
putative explanatory causes for homosexuality: psychoanalytic, endocrinal,
and sociobiological. Again, my main intent will be expository rather than
critical. At each point I shall see what light is thrown on the 'homosexuality
as sickness' question by comparison of the scientific claims with the philosophical models. What I shall argue is that, given different scientific claims
and different philosophical models, we get different answers to our main
question. This, I shall suggest is the reason why we get such a controversy
over the medical status of homosexuality.

On the point of terminology, as will be obvious, I use the term 'sickness'
in a broad sense to cover 'disease', 'illness', or anything which might be
thought to be the opposite of health.

10.1. TWO MODELS OF HEALTH AND SICKNESS

There are, I believe, two different approaches to the health/sickness question currently supported by philosophers and health care professionals: approaches which, for obvious reasons, I shall label the *naturalist* and *normativist* models respectively. Both approaches see the healthy person as someone whose parts are working in a satisfactory manner, that is to say, someone whose parts 'function properly'. The sick person, therefore, is someone whose parts are not working properly, who does not function adequately. Where the naturalist and normativist models differ is over precisely what constitutes proper working or functioning. The naturalist, in recent years most particularly and articulately Christopher Boorse (1975), (1976a), (1977), tries to tie functioning in as far as possible to the biological. For him/her, a body part functions properly if it contributes to survival and reproduction – obviously it need not contribute to indefinite survival and reproduction, but rather to that which is typical of the species. In other words, a body part is healthy if it lives up to the 'species design'. If it does not, then it is 'diseased' and the bearer has a 'disease'. A man with mumps in his testicles, therefore, probably has reduced reproduction chances and, hence, has a disease. Similarly for a woman dying of leukemia.

Notice that for Boorse, the concept of disease has no value connotations *per se*. It simply refers to something breaking down and implies nothing about whether the process of effect is unpleasant or not desired by the individual involved. "On our view disease judgements are value-neutral ... their recognition is a matter of natural science, not evaluative decision" (Boorse, 1977, p. 534). But obviously sometimes, often, we do want to express regret at or dislike of a disease. If a person has lung cancer then he/she is unhappy. In order to capture this facet of ill-health, Boorse draws a distinction between disease and 'illness'. Illnesses are those members of the sub-class of diseases that we do not want. Therefore, we can and do introduce values through this notion of illness. In short, lung cancer is a disease and, as such, is a state – no more, no less. However, people with lung cancer are also ill, meaning that they have something they would rather not have. Conversely, if one had something reducing reproductive ability but were unconcerned – say one were a Catholic father of ten with blockages leading to a *de facto* vasectomy – one would have disease but no illness.

In contrast to the naturalist, the normativist sees proper working or functioning as something always involving norms or values (Margolis, 1976, 1980; Engelhardt, 1975, 1976). For humans, proper functioning is a cultural phenomenon which is bound up with fitting or meshing into what we in society find worthwhile. In other words, the healthy person is the person whose body and attributes enables them to participate fully and happily in society and everyday life. As one supporter of this position states: " . . . the norms of health and disease tend to correspond — often in a disputatious way — with putative norms of happiness and well-being . . . To the extent that this occurs, it becomes difficult to treat the norms of medicine as altogether independent of ideologies, prevailing in different societies" (Margolis, 1976, p. 345).

Where then lies the distinction between disease and illness for the normativist? Unlike the naturalist, the normativist judges them both as being unpleasant or unwanted. The difference is more one of causal relationship: a disease is something one has which makes one ill or perhaps even more intimately, an illness is simply a disease one is physically aware of. The above-quoted supporter states that "illness is reflexively palpable disease" (ibid.). Another writer puts matters thus: "We identify illness by virtue of our experience of them as physically or psychologically disagreeable, distasteful, unpleasant, deforming — by virtue of some form of suffering or pathos due to the malfunctioning of our bodies or our minds. We identify disease states as constellations of observables related through a disease explanation of a state of being ill" (Engelhardt, 1976, p. 259).

These then are the two current models of health and sickness with which we shall work. For the naturalist there is a crucial distinction between disease and illness. Disease is a state reducing reproductive/survival chances. Illness is a disease one does not want: something which makes one unhappy. For the normativist both disease and illness make one unhappy — the distinction is rather between cause and effect, phenomenon and feelings. Consequently, one could not possibly want or live comfortably with either a disease or an illness. Conversely however, even if some phenomenon reduces either reproductive or survival chances, if one does not very much care, one has neither disease nor illness.

Given our two models, we must now turn to what is known, or at least claimed, about homosexuality. But as we do so, let me note one area of controversy that I am avoiding entirely. In recent years, the notions of mental health and mental illness have come under strong attack, both as incoherent and as socially dangerous (especially from Szasz (1961)). Clearly this attack has to be defused to open the way for a positive judgement about the disease/illness status of homosexuality, because (whether or not there be a physical cause) homosexual orientation as such is a 'mental' phenomenon.

However, at this point I shall not get side-tracked into a discussion of the logical/ethical status of mental illness. I sympathize very much with the intentions of the critics – I hate the way that so many today refuse to judge behaviour because of popular half-baked psychological theses. But as a general charge, I think the attack against the concept of mental illness fails, and that others have already adequately shown this. Hence, presupposing the defenses, I shall assume that one can legitimately ask questions about mental health and mental· disease/illness (see Margolis, 1966; Macklin, 1972, 1973; Moore, 1975a, 1975b; Klerman, 1977; Flew, 1973; Brown, 1977).

10.2. THE EMPIRICAL FACTS ABOUT HOMOSEXUALITY

In this paper, for my direct empirical information I shall rely heavily on the recent Kinsey Insitute-endorsed study of homosexuals (and heterosexuals) drawn from the San Francisco area (Bell and Weinberg, 1978). I realize that this study will hardly be the last word on homosexuals, their perceptions of themselves, their life-styles, and so forth. But it does seem to me to be the most comprehensive and best evidence that we have so far: the authors of the study, Alan Bell and Martin Weinberg, gave extensive questionnaires to 972 homosexuals and 477 heterosexuals. (There was also a pilot study run in Chicago.) Since we are already in a situation where people are making judgements about health and illness with respect to homosexuality, it is not as if I am being that premature in thus turning to the Bell/Weinberg report. I shall, however, look briefly at other work. With an eye to our models, there seem to be two kinds of questions we want answered. The first question is: How does a homosexual orientation affect people's functioning, particularly their biological functioning? The second kind of question we want answered is: How do people feel about themselves? Generally speaking, is one happy as a homosexual? Would one be happier were one a heterosexual? Let us take up these questions in turn.[2]

Unfortunately, dead people tend not to complete questionnaires, so there is no direct information in the Bell/Weinberg report, nor anywhere else to the best of my knowledge, on the survival chances of homosexuals as compared to heterosexuals. However, there is pertinent information in the report on the subject of reproduction. As might be expected, it really does seem that having a homosexual rather than heterosexual orientation reduces one's chances of having offspring: in biological language, one has 'reduced fitness'. The chances of getting married if one is a homosexual is very much less than if one is a heterosexual, and even if one does get married, one tends to have fewer

children. Roughly speaking, summarizing much data, only about one homosexual in four married, as opposed to three heterosexuals in four, and of
those that were married, half the homosexuals had no children at all, whereas
two out of three married heterosexuals had children (Bell and Weinberg,
1978, Tables 17.1, 19.3). Moreover, one finds a similar imbalance if one looks
at the total number of children homosexuals and heterosexuals parented.
Of course, there are obvious questions about gaps in this information. For
instance, one certainly does not have to get married to have children. But,
in the absence of evidence to the contrary, one can assume that the missing
information would not seriously distort the conclusion that homosexuals
have fewer children than heterosexuals. Indeed, that additional evidence
we do have points the other way. For example, while they were married
homosexuals had significantly less intercourse (with their spouses) than
heterosexuals (ibid., Tables 17.5, 19.6).

Turning to our second question about feelings of happiness, the Bell/
Weinberg findings do not point unambiguously in one direction. Take first
the question about how homosexuals feel about being homosexual. Are they
happy with their orientation? Or would they wish (have they wished) to
change? (No comparable questions were asked of heterosexuals. I assume the
authors believed that, although some heterosexuals have sexual problems,
they do not wish to become homosexual.) The answers are mixed, although
on balance they strongly support the claim that most homosexuals are
reasonably satisfied with their sexual orientation. Indeed, over three-quarters
of the respondents reported 'none' or 'very little' regret over their homosexual
inclinations (ibid., Table 12). Amongst those who did feel a sense of regret,
societal rejection was the main cause (about 50%), followed by the fact (or
prospect) of not having children (about 25%). In their conclusion on these
findings, the authors (Bell and Weinberg) point out that they find far more
satisfied homosexuals than do surveys restricted to people in treatment, and
they suggest that such latter surveys suffer from the fallacy of biased statistics.
Many homosexuals in treatment are there precisely because they cannot
handle their sexual orientation. I suppose a critic might object that a sample
drawn from the San Francisco population is not that representative either.
If one can live openly as a homosexual one might adjust a lot more readily
to one's orientation than if one is in a society like Guelph, Ontario, where
it is politic to conceal it.

Next, let us look at direct happiness perceptions ('psychological adjustment'). The most obvious, and not necessarily the worst way of finding out
how people feel is to ask them. When the question was put to people, the

Kinsey researchers found that most people were pretty satisfied with their lot, and that there was not indeed a great deal of difference between homosexuals and heterosexuals (ibid., Table 21.3). At least 80% of the respondents, of whatever sexual orientation, reported that they were either 'pretty happy' or 'very happy'. Bell and Weinberg drew up a five-point typology of homosexuals. 'Close-coupled', those living reasonably monogamously with a partner; 'Open-coupled', with a partner but quite promiscuous; 'Functional', unattached, promiscuous, but satisfied; 'Dysfunctional', unattached, promiscuous, and not satisfied; and 'Asexual', not really that turned on by sex. As might be expected, in terms of this typology, Close-Coupleds and Functionals tended to score highest on the happiness scale, and Asexuals and Dysfunctionals lowest. One interesting finding was that, "although the Asexuals and the Dysfunctionals were less happy than the heterosexual men, the Close-Coupleds tended to report even more happiness than those in the heterosexual group" (ibid., p. 199).

A critic is going to object that a major problem with questions like "Are you happy?" is that people simply do not always tell the truth, either about how they feel nor or how they have felt in the past. This is simply a function of the fact that we all have a remarkable capacity for self-deception. I suspect that there is some force in this objection. Although one must certainly take people's self-perceptions seriously, it is undoubtedly true that people do deceive themselves. And luckily, many of us have a great capacity to blank out unhappy memories: "Oh yes, it was a wonderful holiday. So relaxing!" There is, however, one question which the researchers asked which perhaps helps us to edge a little closer to finding out how people really feel about themselves. This was to do with suicide — about whether people had actually attempted it or thought seriously about it. It is certainly not a perfect gauge of mental happiness. Thankfully people who have been suicidal at one point do not always remain so. But the topic of suicide would seem to tell us something about how people feel about themselves, how stable they are, how they adjust to life, how happy they are, and how content they are with their lot. The difference between homosexuals and heterosexuals, particularly homosexual and heterosexual men, are really quite staggering. Homosexuals are far more likely than are heterosexuals to have attempted or seriously considered suicide (ibid., Table 21.12). Nearly 40% of the male homosexuals have tried or seriously contemplated suicide, whereas less than 15% of the male heterosexuals have tried or seriously contemplated suicide. The figures become even more discrepant if one considered only those who have actually tried suicide: a six-to-one imbalance. The differences are not quite as striking in the case of homosexual/heterosexual women, but they are there nevertheless. Of course, one might wonder how representative these figures are, especially given that San Francisco has one of the highest suicide rates in the U.S., but, in fact, the pilot study in Chicago found much the same suicide-attempt rate for homosexuals. Also plausibly, one might suggest that the high homosexual ratio in San Francisco is not independent of a high suicide/suicide-attempt rate.

One wants to know to what extent the high figures for homosexuals are a function of their homosexuality, and whether this be internal (e.g., personal dissatisfaction with their homosexuality) or external (e.g., societal disapproval of homosexuals). Summarizing pertinent data, although not all of the suicide attempts or thoughts are related to homosexuality, something of the order of half are (ibid., Tables 21.13, 21.16, 21.17, 21.18). Putting this information together with everything else, we can conclude that at least one homosexual in five is troubled by homosexuality to the extent of trying or seriously thinking about suicide. And at least half of these are really having severe personal problems over their homosexuality, where these problems not simply a function of societal disapproval: they would like to be heterosexual so they could enter into a monogamous marriage with children, and so forth. One interesting fact is that suicide attempts tend to be the province of the young, for both homosexuals and heterosexuals (ibid., Table 21.15). The obvious implication is that whatever conclusions one might draw about disease and illness may well be more applicable to younger people than to older people.

But what conclusions might we want to draw from all of these facts and figures? Fairly obviously, if we are prepared to take the Bell/Weinberg findings seriously (which I am), I think we would want to say that most homosexuals are pretty content with their lot. Some indeed, particularly Close-Coupleds and Functionals, are very positive about their homosexuality, apparently being at least as satisfied, if not happier than comparable heterosexuals. There really seems no reason to pretend otherwise. On the other hand, there are homosexuals who have trouble accepting their sexual orientation, and there are homosexuals (I would imagine much the same group) who are not very happy. Moreover, many homosexuals have gone through crises of one kind or another which have driven them to serious thoughts of, or even attempts at, suicide. One simply cannot deny this fact, or that we are looking at a minority of the order of 10–20%; although admittedly, this might in part be a transitory phase of development, say pre-25, which is then followed by a happier maturity.

Does a conclusion like this, that most homosexuals are reasonably happy about their orientation but that there is an undoubted minority who are not, mesh fairly well with the findings of other empirical and like inquirers? Although there has been extensive testing of homosexuals and supposedly comparable heterosexuals in order to answer this question – particularly testing based on various types of personality questionnaires – it is not really possible to given an absolutely unequivocal answer. Indeed, to put the matter bluntly, one can find totally conflicting findings about how homosexuals feel about themselves and how their orientation affects their feelings. Some researchers have concluded that homosexuals are an extremely unhappy set of people and that their low feelings about themselves are direct functions of their homosexuality; others, to the contrary, have found far less difference between homosexuals and heterosexuals. Given such disparate claims, one starts to look for obvious causes for the inconsistencies, such as biased selection of subject groups – and one does not have to look far. For instance, one of the most-often quoted studies, purportedly showing that homosexuals are, on average, an unhappy group of people (or rather that male homosexuals are unhappy men), was based exclusively on men in trouble with the military – a notoriously distorted sample (Doidge and Holtzman, 1960). Moreover, even then the researchers had to take a sub-group to get the results they desired! Faced with work of this quality, one is hesitant to put any reliance in its findings. However, surveying the whole field, paying particular attention to that work which seems least flawed, it does seem generally fair to conclude that the consensus of opinion is in line with the Bell/Weinberg findings. Most homosexuals are reasonably content with their lot; a minority are not. (Fortunately, I am freed from the necessity of surveying all the work on homosexual happiness feelings by the existence of a magnificent critical review of the whole field by Gonsiorek (1977). His conclusions are the same as mine.)

So how do these empirical findings fit in with our models of health, disease, and illness? Taking first the naturalist position promulgated by Boorse and concentrating on the negative notions (i.e., disease and illness), we have seen nothing directly about how homosexuality affects survival prospects. Only if we link attempted suicide with successful suicide, can we suggest that homosexuality may, in some respects, reduce survival chances (although this may not be such a ridiculous link). However, we have got reasonably strong evidence that being a homosexual on average reduces reproduction, and this clearly seems to be a function of homosexuality itself, rather than something else. Hence, on one prong of Boorse's criterion, homosexuality must be judged a disease. (See Stengel (1974) on suicide and attempted suicide.)

However, illness is another matter. Some homosexuals, indeed most homo-sexuals, cannot be judged ill by Boorse's criterion. They are at least as satisfied with their lot, specifically with their sexual orientation, as are heterosexuals. Nevertheless, I do think we have to allow that by the naturalist criterion, a small minority of homosexuals are ill, and that this illness must be laid at the feet of their homosexuality. There are, for instance, some who would really like to be heterosexual, to marry, and to have children and, because of their orientation, they cannot and this makes them unhappy. These people are ill, as perhaps are others who at various times in their lives are driven to the brink because of their sexual orientation.

The normativist conclusion overlaps in part with the naturalist conclusion, but not entirely. Those homosexuals that the naturalist would judge to be *both* diseased and ill (because of their homosexuality) would seem to be judged both diseased and ill (because of their homosexuality) by the nor-mativists. Certainly, these are people who are not much enjoying life because of their sexual orientation, and this all seems to fit the normativists' criteria for disease and illness. However, the normativist parts company with the naturalist's claim that, judgements of illness apart, homosexuality generally is a disease. For Boorse, homosexuality is a disease because it reduces biological fitness. For the normativist, whether loss of biological fitness is a disease is a contingent matter, dependent upon whether such loss makes the loser in any way regretful or unhappy. And apparently, since many homosexuals are happy in their homosexuality, the normativist would have no reason to judge them diseased (or ill). One can put matters this way. Boorse would say of the integrated happy homosexual that he/she had a disease but was not thereby ill (see Boorse himself, 1975). The normativist would deny both disease and illness. In Appendix 1, I have drawn up a matrix trying to incorporate all the results I get in this essay. Hopefully, it will help the reader to keep things reasonably clear!

In concluding this section, a number of points of clarification and qualifi-cation seem appropriate. First, even if one does judge (some) homosexuals diseased/ill on the basis of the empirical data, this does not imply that they are diseased/ill all of their lives. If anything, the data seems to imply that homosexuality-as-disease/illness is more of a young person's problem, perhaps giving the lie to the popular notion about the tragedy of the aging homosexual – a sentiment which Bell and Weinberg (1978) endorse. Whether spontane-ously or through human intervention, the possibility that homosexuality-as-disease/illness will vanish is certainly not barred.

.Second, if one talks in terms of 'cure' at this point, and the argument does seem to imply that such talk is appropriate for some, note that such cure does not necessarily entail changing a person's sexual orientation, even if this be at all possible. The way to dissolve homosexuality-as-disease/illness may be to come to accept one's sexual orientation, and to appreciate and cherish it for its own values and virtues. Third, related to this last point, total cure might well (undoubtedly will?) involve the heterosexual majority as well. If the majority stop thinking of homosexuality as a handicap and as something unpleasant, and if they stop hating homosexuals, then if nothing else, we shall get a raise in the self-image of presently troubled homosexuals. This is not to deny that the evidence does seem to be that there is more to the problem than societal attitudes. As noted for instance, some homosexuals dislike their homosexuality because they want to be heterosexual, get married, and have children. Admittedly society endorses the having of children, but the happiness of child-rearing transcends societal approval.

The final point is directed against those who might be inclined to say that homosexual happiness cannot, in any way, be compared to heterosexual happiness — it is obviously lower and, therefore, even your integrated homosexual is sick compared to your average heterosexual. All one can say in reply to an objection like this is that there is not the slightest bit of evidence for such an *ad hoc* assumption. The closest parallel would perhaps be the case of a child with Down's Syndrome. We would want to say that such children have both disease and illness, judged by either naturalist or normativist criteria, even though people with Down's Syndrome tend to have exceptionally cheerful and loving natures. I take it that at this point we have to introduce some element of paternalism, saying that if only Down's Syndrome sufferers knew better, they would be unhappy with their lot. But the reason why paternalism is justified here is that people with Down's Syndrome cannot know better — they are retarded. However, no similar case can be made against homosexuals — if anything, the evidence is that they are more intelligent than heterosexuals (Weinrich, 1976). Nor do homosexuals seem lacking in other human virtues which might make us feel permitted to judge for others as, for instance, in the case of the intelligent paranoid schizophrenic. Hence, in the absence of contrary evidence, the objection fails.

It might be felt at this point that there is really no need to take the discussion any further. One might feel that judgements of disease and illness (although, of course, not necessarily of cure) can and should be made directly and exclusively on the basis of empirical studies. If a homosexual has more or fewer children, is more or less happy, that is all we need to know. For our present purposes, there is no necessity to inquire after causes. However, I would suggest to the contrary that it does, in fact, seem worthwhile looking into causes. Obviously the Bell/Weinberg study, although the best of its kind, is somewhat limited in value. Moreover, the various other empirical studies hardly provide the basis for a definitive answer. Hence, we are not yet in such a position of strength that we can turn away possible avenues of help. Furthermore, a look at causal claims might throw a different light on some of the answers to the empirical questions, although I hasten to add, not necessarily a truer light.

For these and like reasons, let us turn to causes. In what follows, for the sake of argument I shall be assuming that the putative causal explanations are well established. The question then is the extent to which such explanations force us to modify judgements about homosexual mental health based purely on empirical claims. I do emphasize that mine is an initial investigation – I am trying to find out where particular suggestions lead us. In this essay I am not judging the truth or falsity of the suggestions, although *en passant* and in conclusion, I shall make a few remarks pertinent to this matter, as I have already done in previous essays: specifically in Essay 8, with respect to sociobiological explanations, and in Essay 9, with respect to endocrinal explanations.

10.3. PSYCHOANALYTIC CAUSAL EXPLANATIONS

I find two main explanations of homosexuality in the psychoanalytic litera-ture. One I shall term the 'classical Freudian' explanation; the other explana-tion stems from Freud's ideas but was developed in part in reaction to Freud and is, for fairly obvious reasons, called the 'adaptationist' or 'phobic' theory.

Freud (1905) argues that we are all essentially, physiologically and psy-chologically, bisexual – with elements of male and female mingled in our nature. In the case of biological males (which was Freud's paradigm), psycho-logically speaking we start out with the male side predominant, as we are attracted to mother. Then the female side comes to the fore, as we turn narcissistically towards our own bodies. And then comes the swing back to the male side, first to mother again, and then usually with a successful resolution of the Oedipus Complex, to other females at adolescence. Females take a comparable path, although as every critic has been happy and eager to point out, Freud gives female psycho-sexual development little study in its own right, and that there it certainly seems to show a sexist bias. Fortunately, Freud's problems are not our problems here (but see Essay 9).

For Freud, therefore, a person is not really turned into a homosexual –

rather, it is more a question of the homosexual side of our nature coming to the front and pushing back or out our heterosexual side (Fisher and Greenberg, 1977). This can happen for a number of reasons. One can be born innately having a strong homosexual disposition. One might, for various environmental reasons, just stay in the homosexual phase of childhood. In boys, this could be because one had a very overpowering and affectionate father. Freud (1905) speculated that the reason why homosexually was so common in Ancient Greece was because male slaves attended and reared young boys. Or one might later have trouble breaking from a dominant and over-affectionate mother, and then coming up against the universal incest taboo (Freud, 1913), one rejects females in favour of males. Freud said little about lesbianism, but his few comments show that he thought analogous innate or environmental forces might keep a person in a usually pre-adult state of psycho-sexual development (Freud, 1920).

Consequently, for Freud homosexuality was not so much 'abnormal', but more a question of arrested development or regression to an earlier stage. This means that, in his language, it is a 'perversion' rather than a 'neurosis', which latter involves repression of memories, ideas, and so forth, from full consciousness. Hence, the method of analysis is inappropriate, since this is designed to bring unconscious elements to the front. Freud was, therefore, not very sanguine about curing homosexuality, in the sense of changing one predominantly homosexual into one predominantly heterosexual. Indeed, as he explained in his famous 'Letter to an American Mother', he did not even think that homosexuality generally is an illness. (The full letter is given in Appendix 2. It gives a perfect summary of Freud's main position.)

The question we want to ask is whether, if one accepts Freud's position as true (please note the hypothetical), we would want to modify or in any way alter the conclusions arrived at in the last section. Freud himself seems to imply that if one regards homosexuality as a case of arrested development (which he does), then this means it cannot be an illness. However, this seems not to follow. In girls, if a piece of one of their sex chromosomes is missing, there is a failure to develop sexually at puberty. We would (and do) certainly want to classify this phenomenon, Turner's Syndrome, as a disease/illness (Levitan and Montague, 1977). Therefore, analogously, such judgements seem not to be ruled out in the case of homosexuality. But, of course, this is not to say conversely that even if homosexuality is arrested development, that it is thereby either a disease or an illness.

It would, in fact, seem that accepting this classical Freudian position would not really much affect judgements made on the empirical evidence alone. The Freudian position implies nothing new about survival or reproduction, nor about the happiness or unhappiness of homosexuals. Perhaps, indeed, it does imply (as certainly Freud's letter assumes) that there is no reason why a homosexual cannot be perfectly happy and content with his/her lot — that such happiness is genuine. Some people are content to stay in or go back to a phase of childhood development. This being so, both naturalist and normativist can agree with Freud that many homosexuals are not ill at all. This is not to deny the other conclusions we arrived at, for instance, that Boorse would judge homosexuality a disease, or that both normativists and naturalists could subscribe to Freudian views and think some homosexuals ill.

In fact, as we shall now see, inasmuch as Freud subscribed to a version of the phobic theory — which certainly grows out of his ideas — he too would probably have judged some homosexuals sick. Remember that, for Freud, one thing which can drive someone back to a homosexual phase, that is which can lead to an incomplete resolution of the Oedipal strife, is an overly close mother. Another thing, one which often accompanies such a mother, is a hostile threatening father. The developing boy sees father as a rival threat to his love of mother. Since father is stronger than the son, the son does the safest thing, which is to take himself out of the heterosexual running! Often this is all bound up with the fact that, at an earlier stage in life when the boy had discovered his own penis (the 'phallic phase'), there was also the discovery that females do not have penises — in the boy's view, they are castrated. Hence, the threatened son fears that the father will geld him, if he continues to look towards the female side of the human race (Freud, 1905).

These sorts of ideas have been taken up by a number of analysts and developed into the adaptationist or phobic theory (Bieber *et al.*, 1962; Bieber, 1965; Kardiner *et al.*, 1959a, 1959b, 1959c, 1959d; Ovesey, 1954, 1956; Rado, 1940, 1949; Salzman, 1965). Breaking from Freud, they see our natural state as heterosexual — we are not constitutionally bisexual. Normally a boy (they are good Freudians in that they continue to take males as the paradigm!) will grow up to be attracted to females, but certain things — fears — may deflect him into homosexuality. Homosexuality is, therefore, an adaptive move in the fact of perceived threats. The fears may start at an early stage of life. Seeing the gash caused by the girl's castration, the boy will fear that a like fate awaits him, and on learning of the mechanism of sexual intercourse the boy might hypothesize that this is the main source of danger — that the vagina contains teeth that lie in wait, ready to bite off the penises of would-be copulators. (This line of argument is taking place subconsciously.

A little lesson in gynecology or dentistry is not going to remove the fear.) Again, the fears might start in the way Freud suggested, as a function of an inadequate resolution of the Oedipus Complex, brought about by a dominant mother and hostile father.

But whatever the cause, the homosexual male is deflected from his 'true' heterosexual nature through fear. In a similar way, because of unusual family dynamics, a girl can be deflected from 'true' heterosexuality: she fears the threatening penis, or some such things. Nevertheless, unpleasant though this might be, in the case of either sex, because the phobic theorists think that we are all 'really' heterosexual, unlike Freud such theorists think that cure (i.e., change to heterosexuality) is theoretically and, in many cases, practically possible. Through analysis, one may remove the fear and curb the false adaptation.

Clearly, if one adopts this position, then some revisions in our conclusions based on the Bell/Weinberg report are called for. There is no change in the belief that homosexuals have lower biological fitness, so the naturalists' claim that homosexuality is a disease is unaltered. But serious doubt is thrown on the claim that one can really be homosexual and happy. All homosexuals are walking on something of an emotional tightrope — one is thinking and acting to cover up dreadful, threatening, irrational fears of castration and death. The happiness one can achieve is at best phony, at worst fictitious. Hence, the phobic theorist will be inclined to disregard the claims of homosexuals that they get any measure of real and lasting happiness — his/her theory tells him/her otherwise. Hence, as a naturalist he/she will judge all homosexuals ill (as well as diseased), and as a normativist he/she will judge homosexuals both diseased and ill. Significantly, it is the phobic theorists who are loudest about the claims that homosexuals are sick, deluded people. At the beginning of this essay we encountered the views of Bieber, a leading phobic theorist, on this subject.

We see, therefore, that we get different answers to our main questions according to which psychoanalytic theories we support. In concluding this section, let me make brief reference to one interesting side claim to the phobic theory. Human beings are essentially heterosexual. This gives the phobic theorists something which needs explaining, namely the fact that sometimes heterosexuals (men especially) have fleeting homosexual dreams, fantasies, or even when they are under stress, encounters. The phobic theorists argue that these are not genuine homosexual phenomenon — they are 'pseudo'. In fact, in an important sense, they are not really sexual phenomena at all. Rather, they are defensive, adaptive reactions to personal problems, like losing out on a professorship to a younger rival (Ovesey, 1955a, 1955b, 1965).

The reaction can go one of two ways. Either one fantasizes about getting on top of things. In a case like that of the rival, one might dream about buggering him, thus acting out one's need to dominate, perhaps even kill, him. Or one retreats to safe ground. A typical pattern would be to identify the penis with that organ which is the symbol of security, the maternal breast. In this case, one wants to suck on the penis as on the breast, and subconsciously one draws up the simple equations: 'penis = breast' and 'semen = milk'.

I take it that by definition, a key aspect of this 'pseudo-homosexuality' is that it is a fleeting aspect of a heterosexual's life, although one may need therapy to get over it entirely. I take it also that it is a rather unpleasant frightening phenomenon. Life is tense enough as it is, without losing grips on one's sense of sexual identity. I rather suspect, however, that at this point (if we accept the theorizing), we get an interesting reversal in our analyses. To now, it has been the naturalist who has been more ready than the normativist with the ascription of a disease. Clearly, here the normativist could and would talk of disease and illness — certainly there is the drop in quality of life that the normativist would seek in applying such labels. But could the naturalist even talk of disease here? I doubt it, because by definition we are dealing with people fundamentally and continuously heterosexual. I see no real loss of biological fitness. Hence, unless one commits suicide there is no loss of life or survival prospects: there is no disease. And hence, there can be no illness either. At most, the naturalist seems locked in to regarding pseudo-homosexuality as one of those unpleasant things which happens to us in the course of life, like grief or fear. One is not sick when one grieves a loss or fears a bull, even though one is not happy. Perhaps this is how the naturalist must categorize pseudo-homosexuality.

One final note in this section: There are other putative causal explanations of homosexual orientation which, like psychoanalytic explanations, give a major role to environmental factors, particularly those involving family dynamics. Especially popular today are various social-learning hypotheses. Most of these explanations are like classical Freudianism, seeing homosexuality in the adult as just as 'authentic' as heterosexuality — there is no implication that all homosexuals are 'really' heterosexual. For this reason let me simply suggest that health/sickness judgements for these other environmentalists should and do not differ much from the judgements of the classical Freudian (West, 1967; Churchill, 1967; Gagnon and Simon, 1973).

10.4. ENDOCRINAL CAUSAL EXPLANATIONS OF HOMOSEXUALITY

There have been two main sets of attempts at pinning the causes of homosexual orientation on hormonal levels, especially sex hormonal levels. One set argues that the crucial causal period is during development; the other that it is during adulthood. Initial work is centered on the second set of hypotheses, suggesting that sexual orientation is primarily a function of male/female sex hormone ratios (androgen/estrogen ratios) in the adult. However, empirical studies have not really borne much fruit in this direction, and so although such hypotheses area certainly not discredited entirely, more and more attention is being directed toward the effects of hormones on development (Meyer-Bahlburg, 1977, 1979; Dörner, 1976). For brevity in this section I shall concentrate exclusively on these developmental hypotheses, although I suspect that much (if not all) of what I have to say would apply to all endocrinal explanations.

Fundamentally, what is argued is about as simple as the psychoanalytic theories are complex. According to the chief spokesman of this position, Gunther Dörner in East Germany, the key organ in sexual orientation is the hypothalamus, and the key time is during its formation which is (approximately) from the third to sixth months of fetal development. Dörner argues straightforwardly that, for males, if at this time the androgen/estrogen level ratio surrounding (or in) the fetus is lower than average, the adult will grow up sexually directed towards other males. Conversely, for females, if the androgen/estrogen level ratio is higher than normal, the adult will grow up sexually directed towards other females (Figure 10.1). And that is that. No change or cure is possible after the hypothalamus is fixed, beyond a vague as-yet-unrealized hope of psycho-surgery to alter sexual orientation.[3]

As we have seen, Dörner himself has no doubts that homosexuality is a dreadful disease and that homosexuals are ill. He has all sorts of plans for testing amniotic fluid and checking sex hormone ratios, tampering with them if need be in order to prevent future adults with homosexual orientations. But these are Dörner's views. The question we want answered is to what extent (if at all), in the light of Dörner's scientific hypotheses, must we modify conclusions about homosexuality *qua* disease/illness drawn just on the empirical evidence?

First, there is the question of biological functioning. It will be remembered that Boorse explicated this both in terms of reproduction and of survival. In fact, biologists are concerned only with reproduction (see Essay 4); but really Boorse cannot avoid mention of survival, because otherwise he would have to

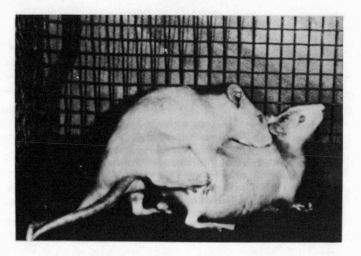

Fig. 10.1. This picture is not quite what it seems. In fact, it is a female rat who is mounting a crouching male rat. By careful hormonal manipulation, Dörner has been able to achieve sexual-behaviour reversal in rats. He thinks one can draw significant analogies between rats and humans, although as can be imagined, many would disagree. (See Dörner, 1976, p. 170.)

say that cancer in an old person is no disease. Now as far as reproduction is concerned, Dörner has nothing new to say. However, most interestingly, he does think his theory has implications about survival. In particular, Dörner (1976) suggests that life-expectancy is a direct function of sex-hormone levels as they have affected the developing hypothalamus (Figure 10.2). The owners of 'male type' hypothalami live shorter lives than the owners of 'female type' hypothalami. This is why, on average, females live ten years longer than males. The implication is, therefore, that male homosexuals will (on average) live longer lives than male heterosexuals, and female homosexuals will live shorter lives than female heterosexuals.

If this claim is true, what happens to an analysis of disease, considered from a Boorsian viewpoint? As far as lesbians are concerned, both reproduction and survival prospects are lowered, and so we have a disease either way. But as far as males are concerned, the model comes apart somewhat. From a reproductive viewpoint male homosexuals are diseased, but from a survival viewpoint male homosexuals are anything but! They are perfectly healthy.

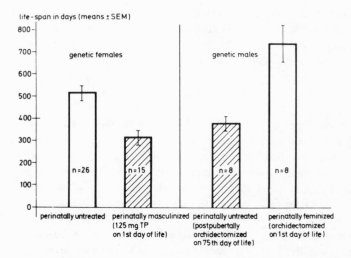

life-span in days (means ± SEM)

genetic females   genetic males

perinatally untreated   perinatally masculinized   perinatally untreated   perinatally feminized
                         (125 mg TP   (postpubertally   (orchidectomized
                         on 1st day of life)   orchidectomized   on 1st day of life)
                                               on 75th day of life)

Fig. 10.2. Continuing his rat analogy, Dörner shows that female rats given male sex hormone (testosterone propionate) live shorter lives than untreated females, whereas males deprived of male sex hormone have longer lives than untreated males. Dörner thinks these findings are relevant to human sexuality (Dörner 1976). One suspects that many sex researchers would disagree, although I know of no direct studies. I imagine testing his claim would pose horrendous difficulties, sorting out the effects of sexual orientation *per se* from other factors, like the different life-styles of homo- and heterosexuals.

Going on to illness, I think a major reason why Dörner thinks homosexuality must be an illness (and a disease too for that matter) is that he believes anything which comes about by accident, for instance, atypical sex hormonal level ratios must be bad for the organism. However, irrespective of whether homosexuality is really a disease, this argument seems not to hold. In the first place, if there is any truth in some of the sociobiological speculations (to be discussed shortly), it is a moot point whether the ratios are all that accidental. But this apart, even though accidents normally cause trouble, judgements of illness must be made solely on the effects. If by accident a child got more growth hormone, taking him from the expected 5′ 6″ to 6′, we would certainly not speak of illness. So similarly we must judge homosexuality.

Staying with Boorse's model, with illness as in the case of disease, we get a certain amount of confusion. Dörner's ideas do not really seem to have any direct implications about how people feel about their homosexuality. Hence,

if (concentrating on reproduction) one thinks of it as a disease, probably
some will have it as an illness also. But not all. However, if by concentrating
on survival, one thinks there is no disease, then there can be no illness;
although no doubt some homosexuals are unhappy with their condition.
Lesbians are easier to handle. By either criterion they have a disease. Some
will be ill also; although this latter is not a judgement based directly on
Dörner's work.

I suppose one might speculate about how homosexuals would feel about
finding out about life-span expectancies! Presumably if anything, this would
tend to make lesbians more miserable, and hence more would be ill, whereas
male homosexuals would be cheered, and hence fewer would be ill. But I
wonder if at this point we are not entering the world of fantasy. My experience is that males including myself do not go around in a state of permanent
gloom because of life expectancy is less than that of women. So I am not sure
how relevant all of this is to be emotional states of homosexuals.

The normativist model does not run into the problems which plague
Boorse's model when it faces Dörner's hypotheses. For the normativist,
reproductive and survival expectancies are secondary to how one feels about
oneself and life in general. Discounting the rather tenuous speculations in the
last paragraph, it really does not seem that, were one to endorse Dörner's
putative causes for homosexual orientation, as a normativist one would much
alter conclusions about disease/illness drawn on the empirical evidence of the
Bell/Weinberg study. We might be prepared to speak of those homosexuals
who are unhappy with their homosexuality as diseased or ill, but there is no
reason to alter the conclusion that for most homosexuals their orientation
is not a disease and, consequently, such homosexuals are no more ill than
comparable heterosexuals. That Dörner himself draws other conclusions
stems not from his causal theorizing, but from his different perception of
the mental state of the average homosexual: a perception which the Bell/
Weinberg study suggests is distorted. One suspects that a major reason Dörner
draws such a gloomy picture of the mental state of homosexuals, is that like
many psychiatrists he is working from a biased sample. He draws his conclusions from homosexuals who are sufficiently troubled to seek help. Happy
homosexuals, like successful murderers, tend not to get into the statistics.

10.5. SOCIOBIOLOGICAL CAUSAL EXPLANATIONS

We come now to the most recent set of explanations of human homosexual
orientation. These are the explanations proposed primarily by biologists who

want to explain human social behaviour as a function of the genes, which have in turn been moulded by the forces of evolution, primarily natural selection (see the last two essays and references). As we know, since selection puts a premium on reproductive efficiency, indeed success in evolution is seen as a success in reproduction, and since *prima facie* homosexual orientation leads to a drop in reproductive efficiency, human homosexuality has attracted much attention from these thinkers, the 'sociobiologists'. We have ourselves encountered two different putative causal explanations from this quarter (Essay 8). Let me reintroduce them, as before in this essay asking whether their truth would require modification of empirically based conclusions about the health/disease/illness status of homosexuality.

The first explanation invokes the so-called phenomenon of 'balanced heterozygote fitness'. Remember that with two alleles A_1 and A_2, and consequently with homozygote genotypes A_1A_1 and A_2A_2 and heterozygote genotype A_1A_2, given the heterozygote fitter than either homozygote, one gets a 'balanced' situation (see Essay 1). In each generation one will get a certain proportion of all three types: A_1A_1, A_1A_2, and A_2A_2. And this situation can continue indefinitely, by virtue of the superior reproductive power of the heterozygote, even though one or other homozygote (say A_2A_2) does not reproduce at all. Applying this model to the homosexuality situation, one simply assumes that there is a gene with two forms, say H and h. If one is an HH homozygote, one is just a normal heterosexual. If one is an hh homozygote, one is a low-reproducing homosexual. And if one is a Hh heterozygote, one is not just a heterosexual but a super-reproducer. Indefinitely, one can expect to get homosexuals in each generation of the human species, even though the homosexuals may contribute little or nothing genetically to each new generation.

Our concern here is not to check whether this model is true or adequate (although see the few brief comments in Essay 8). For the purposes of this discussion we are assuming that the model does hold. Our concern is whether the model's truth leads to qualifications of our empirically-based decisions about the disease/illness status of homosexuality. It seems clear that the model does cause a change of mind for the naturalist, although whether this is a change someone like Boorse would welcome is perhaps another matter. The naturalist sees disease as a failure in proper or adequate functioning. But what is proper or adequate functioning? Not the absolute best in a species; but that which fits into the 'species design'. In other words, that which has brought about and maintained by natural selection. Hence, it is no disease that we cannot synthesize our own vitamin C, because the ability is not something selection left us with. But the whole point of the balanced heterozygote fitness explanation is that the homosexual homozygote is just as much a product of, and just as much maintained by, selection as its heterosexual siblings! In other words, the homosexual is part of the species design, and homosexuality cannot, therefore, be considered a disease. By definition, consequently no homosexual can be made ill by his/her homosexuality, however unhappy it may make him/her feel (see Engelhardt (1976) for a discussion related to the point I am making in this paragraph).

The normativist seems unaffected by the line of reasoning just propounded. That the homosexuality might have been caused by homozygous possession of two alleles seem to make no difference at all to how one feels; except in the indirect and somewhat unpredictable way that knowledge that one's homosexuality is genetic may have on one. Hence, in this case there seems no reason to modify conclusions based on the empirical data.

The second sociobiological model rests on the most exciting mechanism yet proposed by the sociobiologists: kin selection. Inasmuch as relatives reproduce, one reproduces oneself. Hence, a viable biological strategy (kin selection) is to aid relatives to reproduce, even though it may mean reduction of one's own direct reproductive effort. And as we know, some sociobiologists have seen in this kin selection a mechanism for the production of homosexuals (Weinrich, 1976; Wilson, 1978). If for some reason, people would be likely themselves to be low reproducers, or good at aiding relatives (or both), then there might be good biological sense in a condition which would turn one entirely or primarily from attempts at direct personal reproduction: such a condition being a homosexual orientation.

This kin selection explanation has rather interesting implications for the naturalist model of disease/illness. On the surface, the explanation seem to imply that homosexuals will themselves be rather poor reproducers. Furthermore, it rather seems to imply that homosexuals would be the kind of people who (homosexuality apart) would not be very good at reproducing anyway. One proponent of this model has indeed suggested that the reason for kin selection turning one into a homosexual might be a childhood illness (with possible long-term effects) like TB, or just general physical slightness or weakness (Weinrich, 1976). Hence, even though one might not know much about the survival and reproductive chances of homosexuals, in Boorse's sense they seem diseased, and many if not all seem either to be or have been quite ill.

But there are two major points of qualification. First, the proponents suggest that kin selection might have been at work at fullest force amongst our ancestors, when we were all primitive people living closer to nature then we do today. In such a state, it might have paid to switch sexual orientation, given a childhood disease. But in our modern society, biology might not have caught up to today's medicine. A childhood illness may still trigger homosexuality, even though today the child might recover completely, or alternatively, in today's society the fragility which would count against one in a hunter-gatherer society may be no handicap. Hence, we cannot immediately presume that all homosexuals today are either ill or show the effects of such illness. Moreover, as we saw in an earlier essay (8), kin selection might turn one into a homosexual because one has super-characteristics which others do not have. Intelligence is no sickness.

Second, the explanation does not make homosexuality an illness, or even a disease for that matter, whatever the proximate cause. First, there is the fact that the sexual orientation of the homosexual is supposed to have come about in part through selection and this, in itself, means that the naturalist has trouble talking in terms of 'disease' at this point. But secondly, and more importantly, the whole point of the kin selection explanation is that one has reduced reproductive potential and one is perhaps also ill, in the first place. Homosexuality is a biologically adaptive move to *increase* one's reproduction, in the face of this. In other words, from Boorse's viewpoint, homosexuality is no disease – it is a cure! What one is doing is exchanging the uncertain prospects of direct reproduction for the reasonably certain prospects of vicarious reproduction. On this model, therefore, we may be facing people with diseases and illnesses in Boorse's sense, although this is questionable – but homosexuality itself seems no disease in Boorse's sense.

Turning to the normativist's analysis of disease and illness, little more needs to be added to what has just been said in the context of our discussion about Boorse's analysis. There is certainly no especial implication that homosexuality itself is a (normativist) disease or illness, although this is not to deny that it might be an adaptive move in the face of other disease and illness. The one thing about the normativists' position different from the naturalist position that we can say is that, even supposing homosexuality is increasing someone's biological fitness, if the homosexuality makes the bearer unhappy, then we can start to think in terms of disease and illness. I rather think that Boorse has cut off this option. However unhappy homosexuality makes someone, if it does not reduce reproductive (or survival) potential, then it is neither disease nor illness. I am not, of course, suggesting that the kin selection model implies that all homosexuals are unhappy. Indeed, one might go about one's biological business better if one is happy and content. The point is that the model probably does not make the normativist change views based on empirical evidence: most homosexuals are neither diseased nor sick, although a minority may be both, because of their sexual orientation.

10.6. CONCLUSION

With this brief reference to sociobiological explanations of homosexual orientation, we reach the end of our attempt to compare the two major models of disease and illness against the known and speculated facts about homosexuality, empirical and causal. Have we answered our initial question: "Are homosexuals sick?" In fact, we have done so rather too well, because we have come up with a whole set of answers — answers which are far from uniform, as Appendix 1 well shows! However, while in a sense having too many answers is almost as bad as having too few answers, all is far from lost. We now know at least three things we did not know when we set out. First, one of our models of illness and disease, the naturalist model, needs a certain amount of revision if it is to remain a plausible approach to analysing ill-health. The model has led to counter-intuitive results and shown internal strains. If one likes the general approach, it can probably be revised — but revision is necessary. Second, we know that the reason why people differ on the homosexuality and health question is undoubtedly because they bring different assumptions about disease, illness, and homosexuality to their arguments. Now, at least we are in a much better position than previously to sort out what information (or misinformation) is leading to what result. We are closer to finding out why particular people differ. And third, towards an adequate ultimate answer of our own, we do now know what the options are — philosophical and scientific. We have discovered what assumptions to check and evaluate.

This checking and evaluation is, I am afraid, going to need another essay, which simply cannot be squeezed into this collection. However, the reader will have long realized that like all good philosophers I love to rush in and pronounce on scientific matters. Moreover, I simply hate to leave things unsolved, which this essay is in danger of doing with respects to its title-question. Let me therefore conclude on a note which is certainly influenced by philosophical speculation but which undoubtedly goes beyond. If one looks at the matrix I have drawn up (Appendix 1), then despite obvious differences and irregularities certain patterns do start to emerge, especially if one clears up the naturalist position and substitutes the kinds of results that someone like Boorse would probably really like. There is an obvious and expected difference between the naturalist and normativist with respect to the disease-status of homosexuality. The facts are not in real dispute, but the naturalist sees the homosexual as diseased, simply because his/her definition of disease crucially depends on reproduction. The normativist sees no automatic linking of disease and homosexuality. But when it comes to illness, something really important in all eyes because it is undesirable, generally we see naturalist and normativist coming together. Empirical and causal studies both underline the conclusion that generally homosexuality is not an illness, although in some cases we must truly say that people are ill precisely because they are homosexual.

To use a metaphor, the obvious fly in this ointment is the adaptationist/phobic position. According to the theorist who accepts this viewpoint, the homosexual is really sick, however one defines one's various terms. Obviously, therefore, someone must be wrong and equally obviously we cannot assume automatically that, because the phobic theorist goes against the consensus, he/she is the one who is wrong. However, drawing on analyses of previous essays let me leave the reader with this thought. It seems plausible to suggest that any fully adequate science of human behaviour is going to be multi-dimensional, or at least multi-levelled. It is not only going to have to mesh smoothly at the empirical/phenomenal and causal levels, but also most likely it will involve integrated explanations at environmental, hormonal, and quite possibly genetic levels. Specifically in the case of homosexuality, the full explanatory causal picture may well be a composite of several of the putative explanations which have been discussed in this essay, at least explanations exemplifying the various approaches discussed in this essay. All of this being so, there is surely a presumption (no more) in favour of those explanations which already harmonize, particularly if they seem also to fit with the empirical facts as we best know them.

I am not saying absolutely that the adaptationist/phobic theorist is wrong. I am saying that as things stand at present, perhaps the presumption is that such a theorist is not right. In other words, I do argue that given what we know or can best presume, it is unwise and unfair to assume that all homosexuals are sick. The presumption is rather that many homosexuals enjoy full mental health. No doubt, some are troubled because of their homosexuality; but to conclude more than this is to put ideology above science. And, if nothing else, I hope this collection of essays will have shown the reader that one should no more do this than one should try to pretend that science can proceed divorced from ideology. The apprehension of values and facts come together to make a united whole, as the topic of this essay — like those which went before — surely shows.

APPENDIX 1

Matrix comparing models of disease/illness against putative facts about homosexuality, empirical and causal. A certain amount of interpretation has obviously been necessary! Please note I am dealing with averages; hence, the existence of homosexuals with large families does not deny the fact that, on average, homosexuality leads to a reduction in number of offspring.

	Naturalism		Normativism	
	Disease	Illness	Disease	Illness
Kinsey study	All (homosexuals are diseased)	Few	Few	Few
Psychoanalytic Classical Freudian	All	Few	Few	Few
Phobic: homosexuality	All	All	All	All
Pseudo- homosexuality	None	None	All	All
Endocrinal *Male* Survival	None	None	Few	Few
Reproduction	All	Few		
Female Survival	All	Few	Few	Few
Reproduction	All			
Sociobiological Balanced heterozygote fitness	None (All)*	None (Few)*	Few	Few
Kin selection	Some → All**	Few → Many**	Few → Many	Few → Many

* I suspect that this is the result that the normativist really wants.
** But disease/illness is not the result of homosexuality, which is in fact a 'cure'!

APPENDIX 2

Letter from Freud to an American mother worried about her son's homosexuality. (Reprinted in Jones, (1958, 3, pp. 208–9).)

April 9, 1935

Dear Mrs. . . .

I gather from your letter that your son is a homosexual. I am most impressed by the fact that you do not mention this term yourself in your information about him. May I question you, why you avoid it? Homosexuality is assuredly no advantage, but it is nothing to be ashamed of, no vice, no degradation, it cannot be classified as an illness; we consider it to be a variation of the sexual function produced by a certain arrest of sexual development. Many highly respectable individuals of ancient and modern times have been homosexuals, several of the greatest men among them (Plato, Michelangelo, Leonardo da Vinci, etc.). It is a great injustice to persecute homosexuality as a crime, and cruelty too. If you do not believe me, read the books of Havelock Ellis.

By asking me if I can help, you mean, I suppose, if I can abolish homosexuality and make normal heterosexuality take its place. The answer is, in a general way, we cannot promise to achieve it. In a certain number of cases we succeed in developing the blighted germs of heterosexual tendencies which are present in every homosexual, in the majority of cases it is no more possible. It is a question of the quality and the age of the individual. The result of treatment cannot be predicted.

What analysis can do for your son runs in a different line. If he is unhappy, neurotic, torn by conflicts, inhibited in his social life, analysis may bring him harmony, peace of mind, full efficiency, whether he remains a homosexual or gets changed. If you make up your mind he should have analysis with me!! I don't expect you will!! he has to come over to Vienna. I have no intention of leaving here. However, don't neglect to give me your answer.

Sincerely yours with kind wishes,

FREUD

P.S. I did not find it difficult to read your handwriting. Hope you will not find my writing and my English a harder task.

NOTES

1 I have in mind here the kind of distinction discussed in Essay 1, between the upper-level statements of a theory which are used in causal explanation of empirical facts referred to in the lower-level statements.

2 Questions like these, about one's feelings about one's sexual orientation, should not be confused with matters touched on at the end of the last essay, namely the way people feel about their sense of gender identity and associated abilities to carry out roles. There I was talking about people's apparent unhappiness if they do not feel securely identified with male or female — most homosexuals feel just as secure on this matter as do heterosexuals.

3 As hinted in the last essay, until recently theorizing like this was dismissed by sex researchers, particularly those in America. There is undoubtedly evidence that prenatal hormonal levels affect the hypothalamus and can influence subsequent behaviour and thoughts, specifically in directions more typical of the opposite sex. But nearly all denied that hormones could significantly affect adult sexual orientation. Now that some of their subjects are reaching mature adulthood however, I sense that these formerly sceptical researchers are starting to rethink their positions. Perhaps the hormones do count after all! (Money and Schwartz, 1978).

BIBLIOGRAPHY

Adams, M. S. and Neel, J. V. (1967), 'Children of incest', *Pediatrics* **40**, 55–62.

Adkins, E. K. (1980), 'Genes, hormones, sex and gender', in G. W. Barlow and J. Silverberg (eds.), *Sociobiology: Beyond Nature/Nurture*, Westview, Boulder, Col., pp. 385–416.

Alexander, R. D. (1971), 'The search for an evolutionary philosophy', *Proc. Roy. Soc. Victoria Australia* **84**, 99–120.

Alexander, R. D. (1974), 'The evolution of social behaviour', *Ann. Rev. Ecol. Systematics* **5**, 325–84.

Alexander, R. D. (1975), 'The search for a general theory of behaviour', *Behavioral Science* **20** 77–100.

Alexander, R. D. (1979), *Darwinism and Human Affairs*, University of Washington Press, Seattle.

Alexander, R. D. and Sherman, P. W. (1977), 'Local mate competition and patterns in the social insects', *Science* **196**, 494–50.

Allen, E., *et al.* (1975), 'Letter to editor', *New York Review of Books* **22**, 18, 43–4.

Allen, E., *et al.* (1976), 'Sociobiology: another biological determinism', *BioScience* **26**, 182–6.

Allen, E., *et al.* (1977), 'Sociobiology: a new biological determinism', in Sociobiology Study Group of Boston (eds.), *Biology as a Social Weapon*, Burgess, Minneapolis.

Alvarez, L. W., *et al.* (1980), Extraterrestrial cause for the Cretaceous-Tertiary extinction', *Science* **208**, 1095–108.

Aquinas, St. T. (1968), *Summa Theologiae* **43**, *Temperance* (2a 2ae, 141–54), trans. T. Gilby, Blackfriars, London.

Ayala, F. and Kiger, J. (1980), *Modern Genetics*, Addison-Wesley, Reading, Mass.

Ayala, F. J., Tracey, M. L., Barr, L. G., McDonald, J. F., and Perez-Salas, S. (1974), 'Genetic variation in natural populations of five Drosophila species and the hypothesis of the selective neutrality of protein polymorphisms', *Genetics* **77**, 343–84.

Ayala, F. J. and Valentine, J. W. (1979), *Evolving: The Theory and Processes of Organic Evolution*, Benjamin, Cummings, California.

Balian, R., Maynard, R., and Toulouse, G. (1979), *La Matière Mal Condensée* [Ill-condensed Matter], North-Holland, Amsterdam.

Bandura, A. (1969), 'Social-learning theory of identificatory processes', in D. A. Goslin (ed.) *Handbook of Socialization Theory and Research*, Rand McNally, Chicago, pp. 213–62.

Barash, D. P. (1977), *Sociobiology and Behavior*, Elsevier, New York.

Bardwick, J. M. (1971), *Psychology of Women*, Harper and Row, New York.

Barker, A. D. (1969), 'An approach to the theory of natural selection', *Philosophy* **46**, 271–90.

Barlow, G. W. and Silverberg, J. (eds.) (1980), *Sociobiology: Beyond Nature/Nurture?*, Westview, Boulder, Col.

273

Bartsch, F. K. (1978), 'Indications, technique, and limitations of amniocentesis', in J. Scrimgeour (ed.) *Towards the Prevention of Fetal Malformation*, Edinburgh University Press, Edinburgh, pp. 138–51.

Beatty, J. (1978), 'Evolution and the semantic view of theories', unpublished PhD thesis, Indiana University.

Beatty, J. (1980), 'Optimal-design models and the strategy of model building in evolutionary biology', *Phil. Sci.* 47, 532–61.

Beck, L. W. (1969), *A Commentary on Kant's Critique of Practical Reason*, University of Chicago Press, Chicago.

Beckner, M. (1959), *The Biological Way of Thought*, Columbia University Press, New York.

Bell, Alan P. and Weinberg, Martin S. (1978), *Homosexualities – A Study of Diversity among Men and Women*, Simon and Schuster, New York.

Benzer, S. (1962), 'The fine structure of the gene', *Scientific American* 206, 70–84.

Berg, P., *et al.* (1975), 'Summary of statement of the Asilomar conference on recombinant DNA molecules, May 1975', reprinted in C. Grobstein (ed.), *A Double Image of the Double Helix*, Freeman, San Francisco, pp. 113–17.

Bethell, T. (1976), 'Darwin's mistake', *Harper's Magazine* 252, 70–5.

Bieber, I. (1965), 'Clinical aspects of male homosexuality', in J. Marmor, (ed.), *Sexual Inversion: The Multiple Roots of Homosexuality*, Basic Books, New York, pp. 248–67.

Bieber, I. (1973), 'Homosexuality – An adaptive consequence of disorder in psychosexual development, A symposium: Should homosexuality be in the APA nomenclature?', *Am. J. Psychiatry* 130, 1209–11.

Bieber, I., Dain, H. J., Dince, P. R., Drellich, M. G., Grand, H. G., Gundlach, R. H., Kremer, M. W., Rifkin, A. H., Wilbur, C. B., and Bieber, T. B. (1962), *Homosexuality: A Psychoanalytic Study of Male Homosexuals*, Basic Books, New York.

Birch, C. and Abrecht, P. (1975), *Genetics and the Quality of Life*, Pergamon, New South Wales.

Black, M. (1962), *Models and Metaphors*, Cornell University Press, Ithica, N.Y.

Blumberg, B. D., *et al.* (1975), 'The psychological sequelae of abortion performed for a genetic indication', *Am. J. Obstet. Gynecol.* 122, 799–808.

Boorse, C. (1975), 'On the distinction between disease and illness', *Philosophy and Public Affairs* 5, 49–68.

Boorse, C. (1976a), 'What a theory of mental health should be', *J. Theo. Social Behaviour* 6, 61–84.

Boorse, C. (1976b), 'Wright on functions', *Phil. Rev.* 85, 70–86.

Boorse, C. (1977), 'Health as a theoretical concept', *Phil. Sci.* 44, 542–73.

Borgaonkar, D. S. and Shah, S. A. (1974), 'The XYY chromosome male – or syndrome', *Prog. Med. Gen.* 10, 135–222.

Bowler, P. J. (1976), *Fossils and Progress*, Science History Publications, New York.

Bowser-Riley, S. (1978), 'Current problems of amniotic fluid cell culture', in J. Scrimgeour (ed.), *Towards the Prevention of Fetal Malformation*, Edinburgh University Press, Edinburgh, pp. 157–64.

Braithwaite, R. B. (1953), *Scientific Explanation*, Cambridge University Press, Cambridge.

Brewster, D. (1854), *More Worlds than One: The Creed of the Philosopher and the Hope of the Christian*, Murray, London.

Brock, D. J. H. (1977), 'Biochemical and cytological methods in the diagnosis of neural tube defects', *Progress in Medical Genetics*, N. S. vol. 2, Saunders, Philadelphia, pp. 1–37.

Brock, D. J. H. (1978), 'Screening for neural tube defects', in J. Scrimgeour (ed.), *Towards the Prevention of Fetal Malformation*, Edinburgh University Press, Edinburgh, pp. 37–48.

Brooks-Gunn, J. and Matthews, W. S. (1979), *He and She: How Children Develop their Sex-Role Identity*, Prentice-Hall, Englewood Cliffs.

Brown, R. (1977), 'Physical illness and mental health', *Philosophy and Public Affairs* 7, 17–38.

Burchfield, J. D. (1975), *Lord Kelvin and the Age of the Earth*, Science History Publications, New York.

Burkhardt, R. W. (1977), *The Spirit of System: Lamarck and Evolutionary Biology*, Harvard University Press, Cambridge.

Cain, A. J. (1979), 'Introduction to general discussion [of Gould and Lewontin: Spandrels of San Marco]', *Proc. Roy. Soc., Series B* 205, 599–604.

Callahan, D. (1973), 'The meaning and significance of genetic disease', in B. Hilton *et al.* (eds.), *Ethical Issues in Human Genetics*, Plenum, New York.

Callahan, D. (1978), 'Ethical prerequisites for examining biological research: the case of recombinant DNA', in J. Richards (ed.) *Recombinant DNA*, Academic Press, New York, pp. 135–48.

Campbell, D. T. (1974), 'Evolutionary epistemology', in P. Schilpp (ed.), *The Philosophy of Karl Popper*, Open Court, LaSalle, Ill.

Cannon, H. G. (1958), *Lamarck and Modern Genetics*, Manchester University Press, Manchester.

Caplan, A. (1977), 'Tautology, circularity, and biological theory', *Am. Naturalist* 111, 390–3.

Caplan, A. (ed.) (1978), *The Sociobiology Debate*, Harper and Row, New York.

Carlson, E. O. (1966), *The Gene: A Critical History*, Saunders, Philadelphia.

Carter, G. S. (1951), *Animal Evolution*, Sidgwick and Jackson, London.

Chakrabarty, A. M. (1979), 'Recombinant DNA: areas of potential applications', in D. A. Jackson and S. Stich (eds.), *The Recombinant DNA Debate*, Prentice-Hall, Englewood Cliffs, pp. 56–66.

Chambers, R. (1844), *Vestiges of the Natural History of Creation*, Churchill, London.

Chargaff, E. (1976), 'On the dangers of genetic meddling', *Science* 192, 938–40.

Childs, B. *et al.* (1976a), 'Tay–Sachs screening: motives for participating and knowledge of genetics and probability', *Am. J. Hum. Gen.* 28, 537–45.

Childs, B. *et al.* (1976b), 'Tay–Sachs screening: social and psychological impact', *Am. J. Hum. Gen.* 28, 550–8.

Churchill, F. B. (1968), 'August Weismann and a break from tradition', *J. Hist. Biol.* 1, 92–112.

Churchill, W. (1967), *Homosexual Behavior Among Males; A Cross-Cultural and Cross-Species Investigation*, Hawthorn Books, New York.

Cohen, C. (1979), 'When may research be stopped?', in D. A. Jackson and S. Stich (eds.), *The Recombinant DNA Debate*, Prentice-Hall, Englewood, pp. 203–18.

Cole, J. R. and Cole, S. (1973), *Social Stratification in Science*, University of Chicago Press, Chicago.

Cracraft, J. (1978), 'Science, philosophy, and systematics', *Syst. Zool.* **27**, 213–15.

Crow, J. F. (1973), 'Population perspective', in B. Hilton *et al.* (eds.), *Ethical Issues in Human Genetics*, Plenum Press, New York.

Curtis, R. (1979), 'Biological containment: the construction of safer *E. coli* strains', in D. A. Jackson and S. Stich (eds.), *The Recombinant DNA Debate*, Prentice-Hall, Englewood Cliffs, pp. 67–81.

Darden, L. (1980), 'Review of Michael Ruse's, *Sociobiology: Sense or Nonsense?*', *Isis.* **71**, 653–4.

Darlington, C. D. (1966), *Genetics and Man*, Penguin, Harmondsworth.

Darwin, C. (1859), *On the Origin of Species by Means of Natural Selection*, Murray, London.

Darwin, C. (1868), *The Variation of Animals and Plants under Domestication*, Murray, London.

Darwin, C. (1871), *Descent of Man*, Murray, London.

Darwin, C. (1969), *Autobiography*, ed. N. Barlow, Norton, New York.

Darwin, F. (1887), *The Life and Letters of Charles Darwin, Including an Autobiographical Chapter*, Murray, London.

Davis, B. D. (1977), 'The recombinant DNA scenarios: Andromeda strain, chimera, and Golem', *Am. Sci.* **65**, 547–55.

Davis, B. D. (1979), 'Evolution, epidemiology, and recombinant DNA', in D. A. Jackson and S. Stich (eds.), *The Recombinant DNA Debate*, Prentice-Hall, Englewood Cliffs, pp. 137–54.

Davis, B. D. (1979), 'Review of *Recombinant DNA* by J. Richards', *Am. Sci.* **67**, 717.

Davis, B. D. *et al.* (1967), *Microbiology*, Harper and Row, New York.

Dawkins, R. (1976), *The Selfish Gene*, Oxford University Press, Oxford.

Department of Health, Education and Welfare (DHEW) (1980), 'Guidelines for research involving recombinant DNA molecules', *Federal Register* **45**, 20, 6718–49.

Dickerson, R. E. (1978), 'Chemical evolution and the origin of life', *Scientific American*, September, 70–86.

Dobzhansky, Th. (1937), *Genetics and the Origin of Species*, Columbia University Press, New York.

Dobzhansky, Th. (1951), *Genetics and the Origin of Species*, 3rd edn, Columbia University Press, New York.

Dobzhansky, Th. (1970), *Genetics of the Evolutionary Process*, Columbia University Press, New York.

Dobzhansky, Th., Ayala, F. J., Stebbins, G. L., and Valentine, J. W. (1977), *Evolution*, Freeman, San Francisco.

Dobzhansky, Th. and Parlovsky, O. (1957), 'An experimental study of interaction between genetic drift and natural selection', *Evolution* **11**, 311–19.

Doidge, W., and Holtzman, W. (1960), 'Implications of homosexuality among Air Force Trainees', *J. Consult. Psych.* **24**, 9–13.

Donald, I. (1978), 'Ultrasonography in the diagnosis of fetal malformations', in J. Scrimgeour (ed.), *Towards the Prevention of Fetal Malformation*, Edinburgh University Press, Edinburgh, pp. 123–37.

Dörner, G. (1976), *Hormones and Brain Differentiation*, Elsevier, Amsterdam.

Doyle, A. C. (1902), 'Silver Blaze', in *The Memoirs of Sherlock Holmes*, Appleton, New York.

Dunn, L. C. (1965), *A Short History of Genetics*, McGraw-Hill, New York.

Eiseley, L. (1958), *Darwin's Century*, Doubleday, New York.

Eldredge, N., and Gould, S. J. (1972), 'Punctuated equilibria: an alternative to phyletic gradualism', in T. J. M. Schopf (ed.), *Models in Paleobiology*, Freeman, Cooper, San Francisco.

Engelhardt, H. T. Jr. (1975), 'The concepts of health and disease', in H. T. Engelhardt Jr. and S. F. Spicker (eds.), *Evaluation and Explanation in the Biomedical Sciences*, D. Reidel, Dordrecht.

Engelhardt, H. T. Jr. (1976), 'Ideology and etiology', *J. Med. Phil.* 1, 256–68.

Epstein, C. J. and Golbus, M. S. (1977), 'Prenatal diagnosis of genetic diseases', *Am. Sci.* 55, 703–11.

Epstein, S. E. (1976a), 'Epidemic: the cancer producing society', *Science for the People*, July, 4.

Epstein, S. E. (1976b), 'The political and economic basis of cancer', *Technology Review*, July/August, 35.

Erbe, R. W. (1975), 'Screening for the hemoglobinopathies', in A. Milunsky (ed.), *The Prevention of Genetic Disease and Mental Retardation*, Saunders, Philadelphia, pp. 204–20.

Evans, H. J. (1978), 'Environmental hazards', in J. Scrimgeour (ed.), *Towards the Prevention of Fetal Malformation*, Edinburgh University Press, Edinburgh, pp. 66–81.

Falconer, D. S. (1961), *Introduction to Quantitative Genetics*, Ronald Press, New York.

Farley, J. (1977), *The Spontaneous Generation Controversy from Descartes to Oparin*, Johns Hopkins University Press, Baltimore.

Feinberg, J. (1973), *The Problem of Abortion*, Wadsworth, Belmont.

Feyerabend, P. (1970), 'Consolations for the specialist', in I. Lakatos and A. Musgrave (eds.), *Criticism and the Growth of Knowledge*, Cambridge University Press, Cambridge, pp. 197–230.

Fisher, S., and Greenberg, R. P. (1977), *The Scientific Credibility of Freud's Theories and Therapy*, Basic Books, New York.

Fletcher, J. (1973), 'Parents in genetic counselling: the moral shape of decision-making', in B. Hilton *et al.* (eds.), *Ethical Issues in Human Genetics*, Plenum, New York.

Fletcher, J. C. (1979), 'Ethics and amniocentesis for fetal sex identification', *New Eng. J. Med.* 301, 550–3.

Flew, A. G. N. (1967), *Evolutionary Ethics*, Macmillan, London.

Flew, A. G. N. (1973), *Crime or Disease?* Macmillan, London.

Ford, E. B. (1971), *Ecological Genetics*, 3rd edn, Chapman and Hall, London.

Formal, S. B. (1978), 'The pathogenicity of *Escherichia coli*', in J. Richards (ed.), *Recombinant DNA*, Academic Press, New York, pp. 127–32.

Fox, R. (1971), 'The cultural animal', in J. F. Eisenberg and W. S. Dillon (eds.), *Man and Beast: Comparative Social Behaviour*, Smithsonian Institutions Press, Washington.

Fraser, G. R. and Mayo, O. (eds.) (1975), *Textbook of Human Genetics*, Blackwell, Oxford.

Freter, R. (1979), 'Real and imagined dangers of recombinant DNA technology: the need for expert evaluation', in D. A. Jackson and S. Stich (eds.), *The Recombinant DNA Debate*, Prentice-Hall, Englewood Cliffs.

Freud, S. (1905), 'Three Essays on the Theory of Sexuality', in J. Strachey (ed.) *The Standard Edition of the Complete Psychological Works of Sigmund Freud*, Hogarth, London (1953), vol. 7, pp. 125–243.

Freud, S. (1913), 'Totem and Taboo', in J. Strachey (ed.) *The Standard Edition of the Complete Psychological Works of Sigmund Freud*, Hogarth, London (1953), vol. 13.

Freud, S. (1920), 'The psychogenesis of a case of homosexuality in a woman', in J. Strachey (ed.), *The Standard Edition of the Complete Psychological Works of Sigmund Freud*, Hogarth, London (1955), vol. 18, pp. 145–72.

Freud, S. (1927), 'Some psychological consequences of the anatomical distinction between the sexes', *Int. J. Psychoanalysis* 8, 133–42.

Freud, S. (1932), 'Female sexuality', *Int. J. Psychoanalysis* 13, 281–97.

Freud, S. (1933), 'The psychology of women', in *New Introductory Lectures on Psychoanalysis*, Norton, New York, pp. 153–85.

Frieze, I. H., Parsons, J. E., Johnson, P. B., Ruble, D. N., and Zellman, G. L. (1978), *Women and Sex Roles: A Social Psychological Perspective*, Norton, New York.

Gagnon, J. H. and Simon, W. (1973), *Sexual Conduct: The Social Sources of Human Sexuality*, Aldine, Chicago.

Galjaard, H. (1976), 'European experience with prenatal diagnosis of congenital disease: a survey of 6121 cases', *Cytogenetics and Cell Genetics* 16, 453–67.

Galjaard, H. (1978), 'Early diagnosis and the prevention of genetic disease: molecules and the obstetrician', in J. Scrimgeour (ed.), *Towards the Prevention of Fetal Malformation*, Edinburgh University Press, Edinburgh, pp. 3–18.

Geertz, C. (1980), 'Sociosexology', *New York Review of Books* 26, January 24, 3–4.

George, W. (1964), *Elementary Genetics*, 2nd edn. Macmillan, London.

Gerlovich, J. A. *et al.* (1980), 'Creationism in Iowa', *Science* 208, 1208–11.

Gershon, E. S. *et al.* (1977), 'Genetic studies and biologic strategies in the affective disorders', *Progress in Medical Genetics*, Saunders, Philadelphia, N.S. vol. 2, pp. 101–66.

Ghiselin, M. (1980), 'Review of Michael Ruse's *The Darwinian Revolution: Science Red in Tooth & Claw*', *Systematic Zoology*. 29, 108–13.

Giere, R. (1979), *Understanding Scientific Reasoning*, Holt, Rinehart and Winston, New York.

Gish, D. T. (1972), *Evolution: The Fossils Say No!*, Creation-Life, San Diego.

Glass, B. (1963), 'The relation of the physical sciences to biology – indeterminancy and causality', in B. Baumrin (ed.) *Philosophy of Science: The Delaware Seminar*, vol. 1, Interscience, New York, pp. 223–49.

Glass, N. J. and Cove, A. R. (1978), 'Cost-effectiveness of screening for neural tube defects', in J. Scrimgeour (ed.) *Towards the Prevention of Fetal Malformation*, Edinburgh University Press, Edinburgh, pp. 217–26.

Gold, R. (1973), 'Stop it, you're making me sick! A symposium: Should homosexuality be in the APA nomenclature?', *Am. J. Psychiatry* 130, 1211–12.

Goldschmidt, R. B. (1940), *The Material Basis of Evolution*, Yale University Press, New Haven.

Goldschmidt, R. B. (1952), 'Evolution as viewed by one geneticist', *Am. Sci.* 84, 135.

Gonsiorek, J. (1977), 'Psychological adjustment and homosexuality', *JSAS* (American Psychological Association), MS 1478.

Goosens, W. K. (1978), 'Reduction by molecular genetics', *Phil. Sci.* 45, 73–95.

Goudge, T. A. (1961), *The Ascent of Life*, University of Toronto Press, Toronto.

Gould, S. J. (1973), 'Positive allometry of antlers in the "Irish Elk", *Megaloceros giganteus*', *Nature* 244, 375–6.

Gould, S. J. and Lewontin, R. C. (1979), 'The Spandrels of San Marco and the Panglossian paradigm: a critique of the adaptationist programme', *Proc. Roy. Soc., Series B* 205, 581–98.

Goy, R. W. and Phoenix, C. H. (1972), 'The effects of testosterone propionate administered before birth on the development of behavior in genetic female rhesus monkeys', in C. Sawyer and R. Gorski (eds.), *Steroid Hormones and Brain Function*, University of California Press, Berkeley.

Goy, R. W., Wolf, J. E., and Eisele, S. G. (1977), 'Experimental female hermaphroditism in rhesus monkeys: anatomical and psychological characteristics', in J. Money and H. Musaph (eds.), *Handbook of Sexology*, Elsevier, New York, pp. 139–56.

Green, R. (1972), 'Homosexuality as a mental illness', *Int. J. Psychiatry* 10 (1), 77–128.

Green, R. (1973), 'Should heterosexuality be in the APA nomenclature? A symposium: Should homosexuality be in the APA nomenclature', *Am. J. Psychiatry* 130, 1213–14.

Green, R. (1974), *Sexual Identity Conflict in Children and Adults*, Basic Books, New York.

Grene, M. (1958), 'Two evolutionary theories', *Brit. J. Phil. Sci.* 9, 110–27, 185–93.

Grobstein, C. (1979), *A Double Image of the Double Helix: The Recombinant DNA Debate*, Freeman, San Francisco.

Grünbaum, A. (1977), 'How scientific is psychoanalysis?', in R. Stern, L. Horowitz, and J. Lynes (eds.), *Science and Psychotherapy*, Haven, New York.

Grünbaum, A. (1979), 'Is Freudian psychoanalytic theory pseudo-scientific by Karl Popper's criterion of demarcation?', *Am. Phil. Quart.* 16, 131–41.

Genetic Screening Report (GSR) (1975), *Genetic Screening: Programs, Principles and Research*, National Academy of Science, Washington.

Hagard, S. and Carter, F. (1976), 'Preventing the birth of infants with Down syndrome: a cost benefit analysis', *Brit. Med. J.* 1, 753–6.

Hagard, S., Carter, F., and Milne, R. G. (1976), 'Screening for spina bifida cystica: a cost-benefit analysis', *Brit. J. Prev. Soc. Med.* 30, 40–53.

Halstead, B. (1980), 'Popper: good philosophy, bad science?', *New Scientist* 87, 215–17.

Hamilton, W. D. (1964a), 'The genetical evolution of social behaviour, I', *J. Theor. Biol.* 7, 1–16.

Hamilton, W. D. (1964b), 'The genetical evolution of social behaviour, II', *J. Theor. Biol.* 7, 17–32.

Hanson, N. R. (1958), *Patterns of Discovery*, Cambridge University Press, Cambridge.

Hempel, C. G. (1965), *Aspects of Scientific Explanation*, Free Press, New York.

Hempel, C. G. (1966), *Philosophy of Natural Science*, Prentice-Hall, Englewood Cliffs.

Heston, L. L. and Shields, J. (1968), 'Homosexuality in twins: A family study and a registry study', *Arch. Gen. Psychiatry* 18 (2), 149–60.

Hilton, B. *et al* (eds.) (1973), *Ethical Issues in Human Genetics*, Plenum, New York.

Himmelfarb, G. (1962), *Darwin and the Darwinian Revolution*, Anchor, New York.

Hodge, M. J. S. (1971), 'Lamarck's science of living bodies', *Brit. J. Hist. Sci.* 5, 323–52.

Hofstadter, R. (1959), *Social Darwinism in American Thought*, Braziller, New York.

Holling, C. S. (1964), 'The analysis of complex population processes', *Canadian Entomol.* 96, 335–47.

Holton, G. and Roller, D. H. D. (1958), *Foundations of Modern Physical Science*, Addison-Wesley, Reading, Mass.

Hull, D. L. (1972), 'Reduction in genetics – biology or philosophy?' *Phil. Sci.* 39, 491–9.

Hull, D. L. (1973a), *Darwin and His Critics*, Harvard University Press, Cambridge, Mass.

Hull, D. L. (1973b), 'Reduction in genetics – doing the impossible', in P. Suppes *et al.* (eds.), *Logic, Methodology and Philosophy of Science*, Vol. IV, pp. 619–35.

.Hull, D. L. (1974), *Philosophy of Biological Science*, Prentice-Hall, Englewood Cliffs.

Hull, D. L. (1976), 'Informal aspects of theory reduction', in R. S. Cohen *et al.* (eds.), *PSA 1974*, D. Reidel, Dordrecht.

Hume, D. (1779), *Dialogues Concerning Natural Religion*, reprinted in R. Wollheim (ed.), *Hume on Religion*, Collins, London (1963).

Hutchinson, G. E. (1959), 'A speculative consideration of certain possible forms of sexual selection in man', *Am. Nat.* 93 (869), 81–91.

Huxely, J. (1942), *Evolution: The Modern Synthesis*, Allen and Unwin, London.

Huxley, L. (1900), *Life and Letters of Thomas Henry Huxley*, Murray, London.

Huxley, T. H. (1854–8), 'On natural history, as knowledge, discipline, and power', *Royal Inst. Proc.* 2, 187–95; *Scientific Memoirs* 1, 305–14.

Inglis, B. (1971), '*Poverty and the Industrial Revolution*', Hodder and Stoughton, London.

Isaac, G. L. and McCown, E. R. (1976), *Human Origins: Louis Leakey and the East African Evidence*, Benjamin, Menlo Park, Calif.

Isacks, B. *et al.* (1968), 'Seismology and the new global tectonics', *J. Geophys. Res.* 73, 5855–99; reprinted in A. Cox (ed.), *Plate Tectonics and Geomagnetic Reversals*, Freeman, San Francisco (1973), pp. 358–400.

Jackson, D. A. (1979), 'Principles and applications of recombinant DNA methodology', in D. A. Jackson and S. Stich (eds.), *The Recombinant DNA Debate*, Prentice-Hall, Englewood Cliffs, pp. 39–55.

Jackson, D. A. and Stich, S. P. (1979), *The Recombinant DNA Debate*, Prentice-Hall, Englewood Cliffs.

Jenkin, F. (1867), 'The origin of species', *North Brit. Rev.* 42, 149–71.

Jonas, H. (1976), 'Freedom of scientific inquiry and the public interest., *Hastings Center Report*, August, 15–17.

Jonas, H. (1978), 'Straddling the boundaries of theory and practice: recombinant DNA research as a case of action in the process of inquiry', in J. Richards (ed.), *Recombinant DNA*, Academic Press, New York, pp. 253–72.

Jones, E. (1958), *Sigmund Freud: Life and Work*, Hogarth Press, London.

Judson, H. F. (1979), *The Eighth Day of Creation: Makers of the Revolution in Biology*, Simon and Schuster, New York.

Kaback, M. and O'Brien, J. S. (1973), 'Tay–Sachs: Prototype for prevention of genetic disease', in V. A. McKusick and R. Claiborne (eds.), *Medical Genetics*, HP Publishing, New York, pp. 253–62.

Kaback, M. and Zeiger, R. S. (1973), 'The John F. Kennedy Institute Tay–Sachs Program', in B. Hilton *et al.* (eds.), *Ethical Issues in Human Genetics*, Plenum Press, New York.

Kaback, M. *et al.* (1974), 'Approaches to the control and prevention of Tay-Sachs disease', in A. Steinberg and A. G. Bearn (eds.), *Progress in Medical Genetics*, Grune and Stratton, New York.

Kallmann, F. J. (1952), 'Comparative twin study on the genetic aspects of male homosexuality', *J. Nerv. Ment. Dis.* 115, 283–98.

Kan, Y. W., Golbus, M. S., and Trecartin, T. (1976), 'Prenatal diagnosis of sickle-cell anemia', *New Engl. J. Med.* 294, 1039–40.

Kardiner, A., Karush, A., and Ovesey, L. (1959), 'A methodological study of Freudian theory: I. Basic Concepts', *J. Nervous Mental Disease* 129, 11–19.

Kardiner, A. (1959), 'A methodological study of Freudian Theory: II. The Libido theory', *J. Nervous Mental Disease* 129, 133–43.

Kardiner, A. (1959), 'A methodological study of Freudian theory: III. Narcissism, bisexuality and the dual instinct theory', *J. Nervous Mental Disease* 129, 207–21.

Kardiner, A. (1959), 'A methodological study of Freudian theory: IV. The structural hypothesis, the problem of anxiety, and post-Freudian ego psychology', *J. Nervous Mental Disease* 129, 341–56.

Kettlewell, H. B. D. (1973), *The Evolution of Melanism*, Clarendon Press, Oxford.

Kimura, M. and Ohta, T. (1971), *Theoretical Aspects of Population Genetics*, Princeton University Press, Princeton.

King, J. L. and Jukes, T. H. (1969), 'Non-Darwinian evolution', *Science* 164, 788–98.

Kinsey, A. C., Pomeroy, W. B., and Martin, C. E. (1948), *Sexual Behavior in the Human Male*, Saunders, Philadelphia.

Kinsey, A. C., Pomeroy, W. B., Martin, C. E., and Gebhard, P. H. (1953), *Sexual Behavior in the Human Female*, Saunders, Philadelphia.

Klerman, G. L. (1977), 'Mental illness, the medical model, and psychiatry', *J. Med. Phil.* 2, 220–43.

Knutson, D. C. (ed.) (1980), 'Homosexuality and the law', *J. Homosexuality* 5, 3–160.

Koestler, A. (1971), *The Case of the Midwife Toad*, Hutchinson, London.

Kohlberg, L. (1969), 'Stages and sequence: the cognitive-developmental approach to socialization', in I. A. Goslin (ed.), *Handbook of Socialization Theory and Research*, Rand McNally, Chicago.

Körner, S. (1955), *Kant*, Penguin, Harmondsworth.

Krimsky, S. (1979), 'Review of *The Recombinant DNA Debate*, D. A. Jackson and S. P. Stich (eds.)', *Nature* 282, 179–82.

Kuhn, T. S. (1957), *The Copernican Revolution*, Harvard University Press, Cambridge, Mass.

Kuhn, T. S. (1970), *The Structure of Scientific Revolutions*, 2nd edn., University of Chicago Press, Chicago.

Kushner, S. R. (1978), 'The development and utilization of recombinant DNA technology', in J. Richards (ed.), *Recombinant DNA*, Academic Press, New York, pp. 35–58.

Lack, D. (1947), *Darwin's Finches: An Essay on the General Biological Theory of Evolution*, Cambridge University Press, Cambridge.

Lamarck, J. B. (1809), *Philosophie Zoologique* (Paris); trans. as *Zoological Philosophy* by H. Elliot, Macmillan, London (1914).

Laudan, L. (1971), 'William Whewell on the consilience of inductions', *Monist* 55, 368–91.

Laudan, L. (1977), *Progress and its Problems: Towards a Theory of Scientific Growth*, University of California Press, Berkeley.

Leakey, R. and Lewin, R. (1977), *Origins*, Dutton, New York.
Lee, K. K. (1969), 'Popper's falsifiability and Darwin's natural selection', *Philosophy* **44**, 291–302.
Levitan, M. and Montagu, A. (1977), *Textbook of Human Genetics*, 2nd edn., Oxford University Press, New York.
Lewontin, R. C. (1961), 'Evolution and the theory of games', *J. Theo. Biol.* **1**, 382–403.
Lewontin, R. C. (1969), 'The bases of conflict in biological explanation', *J. Hist. Biol.* **2**, 35–46.
Lewontin, R. C. (1974), *The Genetic Basis of Evolutionary Change*, Columbia University Press, New York.
Lewontin, R. C. (1977), 'Sociobiology – a caricature of Darwinism', in F. Suppe and P. Asquith (eds.), *PSA 1976*, Vol. 2, PSA, Lansing, Mich., pp. 22–31.
Lewontin, R. C. (1978), 'Adaptation', *Sci. Am.* **239** (3), September, 212–30.
Li, C. C. (1955), *Population Genetics*, Chicago University Press, Chicago.
Livingstone, F. B. (1967), *Abnormal Hemoglobins in Human Populations*, Aldine, Chicago.
Livingstone, F. B. (1971), 'Malaria and human polymorphisms', *Ann. Rev. Genetics* **5**, 33–64.
Locke, J. (1959), *An Essay Concerning Human Understanding*, A. C. Fraser (ed.), Dover, New York.
Macbeth, N. (1971), *Darwin Retried*, Gambit, Boston.
Mackie, J. L. (1966), 'The direction of causation', *Phil. Rev.* **75**, 441–66.
Macklin, R. (1972), 'Mental health and mental illness: some problems of definition and concept formation', *Phil. Sci.* **39**, 341–65.
Macklin, R. (1973), 'The medical model in psychotherapy and psychoanalysis', *Comp. Psychiatry* **14**, 49–69.
Magee, B. (1974), *Popper*, Woburn Press, London.
Malinowski, B. (1929), *The Sexual Life of Savages in North-Western Melanesia*, Halcyon House, New York.
Malthus, T. R. (1826), *An Essay on the Principle of Population*, 6th edn., London.
Manser, A. R. (1965), 'The concept of evolution', *Philosophy* **40**, 18–34.
Margolis, J. (1966), *Psychotherapy and Morality: A Study of Two Concepts*, Random House, New York.
Margolis, J. (1976), 'The concept of disease', *J. Med. Phil.* **1**, 238–55.
Margolis, J. (1980), 'The concept of mental illness: a philosophical examination', in B. A. Brody and H. T. Engelhardt Jr. (eds.), *Mental Illness: Law and Public Policy*, D. Reidel, Dordrecht, pp. 3–23.
Marmor, J. (1973), 'Homosexuality and cultural value systems. A symposium: Should homosexuality be in the APA nomenclature?' *Am. J. Psychiatry* **130**, 1208–9.
Masters, W. H. and Johnson, V. E. (1966), *Human Sexual Response*, Little, Brown, Boston.
Maynard Smith, J. (1975), *The Theory of Evolution*, 3rd edn., Penguin, Harmondsworth.
Maynard Smith, J. (1976), 'Evolution and the theory of games', *Am. Sci.* **64**, 41–5.
Mayr, E. (1942), *Systematics and the Origin of Species*, Columbia University Press, New York.
Mayr, E. (1960), 'The emergence of evolutionary novelties', in S. Tax (ed.), *Evolution After Darwin*, Vol. I, University of Chicago Press, Chicago, pp. 349–80.
Mayr, E. (1963), *Animal Species and Evolution*, Belknap, Cambridge, Mass.

Mayr, E. (1969a), *Principle of Systematic Zoology*, McGraw-Hill, New York.

Mayr, E. (1969b), 'Scientific explanation and conceptual framework', *J. Hist. Biol.* 2, 123–8.

Mayr, E. (1972), 'Lamarck revisited', *J. Hist. Biol.* 5, 55–94.

McCall, S. (1976), 'Human needs and the quality of life', in W. R. Shea and J. King-Farlow (eds.), *Values and the Quality of Life*, Science History Publications, New York.

McClearn, G. E. and DeFries, J. C. (1973), *Introduction to Behavioral Genetics*, Freeman, San Francisco.

Mertens, T. R. and Polk, N. C. (1980), 'A comparison of thirteen general genetics textbooks', *Am. Biol. Teacher* 42, 274–85.

Meyer-Bahlburg, H. F. L. (1977), 'Sex hormones and male homosexuality in comparative perspective', *Arch. Sexual Behaviour* 6 (4), 297–325.

Meyer-Bahlburg, H. F. L. (1979), 'Sex hormones and female homosexuality: A critical examination', *Arch. Sexual Behaviour* 8, 101–19.

Michalos, A. C. (1976), 'Measuring the quality of life', in W. R. Shea and J. King-Farlow (eds.), *Values and the Quality of Life*, Science History Publications, New York.

Mikkelsen, M. *et al.* (1978), 'Cost-effectiveness of antenatal screening for chromosome abnormalities', in J. Scrimgeour (ed.), *Towards the Prevention of Fetal Malformation*, Edinburgh University Press, Edinburgh, pp. 209–16.

Mill, J. S. (1865), *An Examination of Sir William Hamilton's Philosophy*, Longmans, Green, London.

Miller, H. (1856), *The Testimony of the Rocks, or Geology in its Bearings on the Two Theologies, Natural and Revealed*, Constable, Edinburgh.

Millett, K. (1970), *Sexual Politics*, Doubleday, Garden City.

Millhauser, M. (1954), 'The scriptural geologists. An episode in the history of opinion', *Osiris* 11, 65–86.

Mills, S. K. and Beatty, J. H. (1979), 'The propensity interpretation of fitness', *Phil. Sci.* 46, 263–86.

Mitchell, J. (1972), *Woman's Estate*, Pantheon, New York.

Mitchell, J. (1974), 'On Freud and the distinction between the sexes', in J. Strouse (ed.), *Women and Analysis*, Grossman, New York, pp. 27–36.

Mivart, S. J. (1870), *Genesis of Species*, Macmillan, London.

Money, J. and Ehrhardt, A. A. (1972), *Man and Woman: Boy and Girl: The Differentiation and Dimorphism of Gender Identity from Conception to Maturity*, Johns Hopkins University Press, Baltimore.

Money, J. and Musaph, H. (eds.) (1977), *Handbook of Sexology*, Elsevier, New York.

Money, J. and Schwartz, M. (1978), 'Biosocial determinants of gender identity differentiation and development', in J. B. Hutchison (ed.), *Biological Determinants of Sexual Behaviour*, Wiley, New York, 765–84.

Moore, G. E. (1903), *Principia Ethica*, Cambridge University Press, Cambridge.

Moore, M. (1975a), 'Some myths about mental illness', *Arch. Gen. Psychiatry* 23, 1483–97.

Moore, M. (1975b), 'Mental illness and responsibility', *Bull. Meninger Clinic* 39, 308–28.

Morley, J. (1971), *Death, Heaven and the Victorians*, University of Pittsburgh Press, Pittsburgh.

Morris, H. M. (ed.) (1974), *Scientific Creationism*, Creation-Life, San Diego.

Morris, S. (1978), 'Darwin and double standard', *Playboy*, August, 108.

Mosedale, S. S. (1978), 'Science corrupted: Victorian biologists consider "The woman question" ', *J. Hist. Biol.* 11, 1–55.

Medical Research Council (MRC) (1978), 'An assessment of the hazards of amniocentesis', *Brit. J. Obstet. Gynaecol.* 85, supplement 2, 1–41.

Muller, H. J. (1950), 'Our load of mutations', *Am. J. Hum. Gen.* 2, 111–76.

Nagel, E. (1961), *The Structure of Science*, Routledge & Kegan Paul, London.

Nagel, E. (1977), 'Teleology revisited', *J. Phil.* 74, 261–301.

Nelson, G. (1978), 'Classification and prediction: a reply to Kitts', *Syst. Zool.* 27, 216–17.

National Registry for Amniocentesis Study Group (NICHD) (1976), 'Midtrimester amniocentesis for prenatal diagnosis: safety and accuracy', *J. Am. Med. Assoc.* 236, 1471–6.

Novick, R. (1978), 'The dangers of unrestricted research: the case of recombinant DNA', in J. Richards (ed.), *Recombinant DNA*, Academic Press, New York, pp. 71–102.

Olby, R. C. (1967), *Origins of Mendelism*, Schocken Books, New York.

Olby, R. C. (1974), *The Path to the Double Helix*, Macmillan, London.

Orians, G. H. (1969), 'On the evolution of mating systems in birds and mammals', *Am. Nat.* 103 (934), 589–603.

Oster, G. F. and Wilson, E. O. (1978), *Caste and Ecology in the Social Insects*, Princeton University Press, Princeton.

Ovesey, L. (1954), 'The homosexual conflict: an adaptational analysis', *Psychiat.* 17, 243–50.

Ovesey, L. (1955a), 'The pseudohomosexual anxiety', *Psychiat.* 18, 17–25.

Ovesey, L. (1955b), 'Pseudohomosexuality, the paranoid mechanism, and paranoia: an adaptational revision of a classical Freudian theory', *Psychiat.* 18, 163–73.

Ovesey, L. (1956), 'Masculine aspirations in women: an adaptational analysis', *Psychiat.* 19, 341–51.

Ovesey, L. (1965), 'Pseudohomosexuality and homosexuality in men: Psychodynamics as a guide to treatment', in J. Marmor (ed.), *Sexual Inversion: The Multiple Roots of Homosexuality*, Basic Books, New York, pp. 211–33.

Owen, R. (1834), 'On the generation of the marsupial animals, with a description of the impregnated uterus of the kangaroo', *Phil. Trans.*, 333–64.

Paley, W (1802), 'Natural Theology', in *Collected Works*, Vol. 4, Rivington, London (1819).

Palmer, R. G. (1980), 'Review of *La Matière Mal Condensée*', *Science* 208 (4447), 1025.

Passarge, E. (1978), 'Screening populations for genetic disease', in J. Scrimgeour (ed.), *Towards the Prevention of Fetal Malformation*, Edinburgh University Press, Edinburgh, pp. 19–36.

Patterson, C. (1978), 'Verifiability in systematics', *Syst. Zool.* 27, 218–21.

Peters, R. H. (1976), 'Tautology in evolution and ecology', *Am. Nat.* 110, 1–12.

Platnick, N. and Gaffney, E. (1977), 'Systematics: A Popperian perspective', *Syst. Zool.* 26, 360–5.

Platnick, N. (1978), 'Evolutionary biology: A Popperian perspective', *Syst. Zool.* 27, 137–41.

Polani, P. E. *et al.* (1979), 'Sixteen years' experience of counselling, diagnosis, and prenatal detection in one genetic centre: progress, results, and problems', *J. Med. Gen.* 16, 166–75.

Polanyi, M. (1968), 'Life's irreducible structure', *Science* 160, 1308–12.

Popper, K. R. (1959), *The Logic of Scientific Discovery*, Hutchinson, London.

Popper, K. R. (1962), *Conjectures and Refutations*, Basic Books, New York.

Popper, K. R. (1970), 'Normal science and its dangers', in I. Lakatos and A. Musgrave (eds.), *Criticism and the Growth of Knowledge*, Cambridge University Press, Cambridge.

Popper, K. R. (1972), *Objective Knowledge: An Evolutionary Approach*, Oxford University Press, Oxford.

Popper, K. R. (1974a), 'Darwinism as a metaphysical research programme', in P. A. Schilpp (ed.), *The Philosophy of Karl Popper*, Open Court, LaSalle, Ill.

Popper, K. R. (1974b), 'Intellectual autobiography', in P. A. Schilpp (ed.), *The Philosophy of Karl Popper*, Open Court, LaSalle, Ill.

Popper, K. R. (1975), 'The rationality of scientific revolutions', in R. Harré (ed.), *Problems of Scientific Revolution*, Oxford University Press, Oxford.

Prakash, S. (1972), 'Origin of reproductive isolation in the absence of apparent genic differentiation in a geographic isolate of *Drosophila pseudoobscura*', *Genetics* 72, 143–55.

Proudfoot, N. J. *et al.* (1980), 'Structure and in vitro transcription of human globin genes', *Science* 209, 1329–35.

Provine, W. B. (1971), *The Origins of Theoretical Population Genetics*, Chicago University Press, Chicago.

Rabbitts, T. H. (1976),'Bacterial cloning of plasmids carrying copies of rabbit globin messenger RNA', *Nature* 260, 221.

Race, R. R. and Sanger, R. (1954), *Blood Groups in Man*, 2nd edn., Blackwell, Oxford.

Rado, S. (1940), 'A critical examination of the concept of bisexuality', *Psychosomatic Med.* 2, 459–67, reprinted in J. Marmor (ed.), *Sexual Inversion: The Multiple Roots of Homosexuality*, Basic Books, New York.

Rado, S. (1949), 'An adaptational view of sexual behaviour', in P. Hock and J. Zubin (eds.), *Psychosexual Development in Health and Disease*, Grune & Stratton, New York, 183–213.

Rainer, J. D. *et al.* (1960), 'Homosexuality and heterosexuality in identical twins', *Psychosomatic Med.* 22, 251–8.

Raper, A. B. (1960), 'Sickling and malaria', *Trans. Roy. Soc. Trop. Med. Hyg.* 54, 503–4.

Rawls, J. (1971), *A Theory of Justice*, Belknap, Cambridge, Mass.

Reilly, P. (1975), 'Genetic screening legislation', in H. Harris and K. Hirschhorn (eds.), *Adv. Human Genetics* 5, 319–76.

Rescher, N. (1978), *Scientific Progress*, University of Pittsburgh Press, Pittsburgh.

Reutlinger, S. and Selowsky, M. (1976), *Malnutrition and Poverty*, Johns Hopkins University Press, Baltimore.

Richards, J. (1978a), *Recombinant DNA: Science, Ethics, and Politics*, Academic Press, New York.

Richards, J. (1978b), 'The limitations of broad moral policies', in J. Richards (ed.), *Recombinant DNA*, Academic Press, New York, pp. 157–76.

Rosenberg, A. (1980), 'Ruse's treatment of the evidence for evolution', *PSA 1980*, 1, 83–93.

Rosenberg, B. and Simon, L. (1979), 'Recombinant DNA: have recent experiments assessed all the risks?' *Nature* 282, 773–4.

Rothwell, N. (1979), *Understanding Genetics*, 2nd edn., Oxford University Press, New York.

Ruse, M. (1970), 'Are there laws in biology?', *Aust. J. Phil.* 48, 234–46.

Ruse, M. (1973a), *The Philosophy of Biology*, Hutchinson, London.

Ruse, M. (1973b), 'The value of analogical models in science', *Dialogue* 12, 246–53.

Ruse, M. (1975a), 'Charles Darwin's theory of evolution: an analysis', *J. Hist. Biol.* 8, 219–41.

Ruse, M. (1975b), 'Darwin's debt to philosophy: an examination of the influence of the philosophical ideas of John F. W. Herschel and William Whewell on the development of Charles Darwin's theory of evolution', *Stud. Hist. Phil. Sci.* 6, 159–81.

Ruse, M. (1976a), 'The philosophy of biology', in W. R. Shea (ed.), *Basic Issues in the Philosophy of Science*, Science History Publications, New York, pp. 74–90.

Ruse, M. (1976b), 'The scientific methodology of William Whewell', *Centaurus* 20, 227–57.

Ruse, M. (1976c), 'Reduction in genetics', in R. S. Cohen *et al.* (eds.), *PSA 1974*, Reidel, Dordrecht, pp. 633–51.

Ruse, M. (1977), 'Is biology different from physics?' in R. Colodny (ed.), *Logic, Laws and Life*, University of Pittsburgh Press, Pittsburgh, pp. 89–127.

Ruse, M. (1978), 'Critical notice of Woodfield *Teleology* and Wright *Teleological Explanations*', *Can. J. Phil.* 8, 191–203.

Ruse, M. (1979a), *The Darwinian Revolution: Science Red in Tooth and Claw*, University of Chicago Press, Chicago.

Ruse, M. (1979b), *Sociobiology: Sense or Nonsense?* D. Reidel, Dordrecht.

Ruse, M. (1980a), 'What kind of revolution occurred in geology?' *PSA 1978*, 2.

Ruse, M. (1980b), 'Ought philosophers consider scientific discovery? A Darwinian case-study', in T. Nickles (ed.), *Scientific Discovery*, D. Reidel, Dordrecht.

Ruse, M. (1980c), 'Philosophical aspects of the Darwinian Revolution', in F. Wilson (ed.), *Pragmatism and Purpose*, University of Toronto Press, Toronto.

Ruse, M. (1980d), 'Charles Darwin and group selection', *Ann. Sci.* 37, 615–31.

Ruse, M. (1981), 'Are there gay genes?' *J. Homosexuality*,

Russell, E. S. (1916), *Form and Function*, Murray, London.

Ryle, G. (1949), *The Concept of Mind*, Hutchinson, London.

Sahlins, M. (1976), *The Use and Abuse of Biology*, University of Michigan, Ann Arbor.

Salmon, W. (1973), *Logic*, 2nd edn., Prentice-Hall, Englewood-Cliffs.

Salzman, L. (1965), '"Latent" homosexuality', in J. Marmor (ed.), *Sexual Inversion: The Multiple Roots of Homosexuality*, Basic Books, New York, pp. 234–47.

Schaffner, K. F. (1967), 'Approaches to reduction', *Phil. Sci.* 34, 137–47.

Schaffner, K. F. (1974), 'The peripherality of reductionism in the development of molecular biology', *J. Hist. Biol.* 7 (1), 111–39.

Schaffner, K. F. (1976), 'Reductionism in biology: prospects and problems', in R. S. Cohen *et al.* (eds.), *PSA 1974*, D. Reidel, Dordrecht, pp. 613–32.

Schaffner, K. F. (1980), 'Theory structure in the biomedical sciences', *J. Med. Phil.* 5, 57–97.
Scheffler, I. (1963), *Anatomy of Inquiry*, Knopf, New York.
Schilpp, P. (ed.) (1974), *The Philosophy of Karl Popper*, Open Court, LaSalle, Ill.
Schindewolf, O. H. (1950), *Grundfragen der Paläontologie*, Schweizerbart, Stuttgart.
Schopf, J. W. (1978), 'The evolution of the earliest cells', *Sci. Am.*, September, 110–38.
Scrimgeour, J. B. (ed.) (1978), *Towards the Prevention of Fetal Malformation*, Edinburgh University Press, Edinburgh.
Scriven, M. (1959), 'Explanation and prediction in evolutionary theory', *Science* 130, 477–82.
Sedgwick, A. (1845), 'Vestiges', *Edinburgh Rev.* 82, 1–85.
Science for the People (SftP) (1978), 'Biological, social and political issues in genetic engineering', in A. M. Chakrabarty, *Genetic Engineering*, CRC Press, West Palm Beach Fa. Reprinted in D. A. Jackson and S. Stich (eds.), *The Recombinant DNA Debate*, Prentice-Hall, Englewood Cliffs (1979), pp. 99–126.
Shepher, J. (1971), 'Mate selection among second generation kibbutz adolescents and adults', *Arch. Sex. Beh.* 1 (4), 293–307.
Shepher, J. (1979), *Incest: The Biological View*, Harvard University Press, Cambridge.
Sheppard, P. M. (1975), *Natural Selection and Heredity*, 4th edn., Hutchinson, London.
Shrader-Frechette, K. S. (1980), *Nuclear Power and Public Policy*. D. Reidel, Dordrecht.
Simpson, G. G. (1944), *Tempo and Mode in Evolution*, Columbia University Press, New York.
Simpson, G. G. (1951), *Horses*, Oxford University Press, Oxford.
Simpson, G. G. (1953), *Major Features of Evolution*, Columbia University Press, New York.
Simpson, N. E. *et al.* (1976), 'Prenatal diagnosis of genetic disease in Canada: report of a collaborative study', *Can. Med. Assoc. J.* 115, 739–48.
Sinsheimer, R. L. (1978), 'The Galilean imperative', in J. Richards (ed.), *Recombinant DNA*, Academic Press, New York, pp. 17–32.
Smart, J. J. C. (1963), *Philosphy and Scientific Realism*, Routledge and Kegan Paul, London.
Smart, J. J. C. (1967), 'Utilitarianism', in P. Edwards (ed.), *Encyclopedia of Philosophy*, Collier-Macmillan, New York.
Socarides, C. W. (1973), 'Homosexuality: Findings derived from 15 years of clinical research. A Symposium: Should homosexuality be in the APA nomenclature?' *Am. J. Psychiatry* 130, 1212–13.
Sommerhoff, G. (1950), *Analytical Biology*, Oxford University Press, London.
Spencer, H. (1893), *Principle of Ethics*, Williams and Norgate, London.
Spitzer, R. L. (1973), 'A proposal about homosexuality and the APA nomenclature. A Symposium: Should homosexuality be in the APA nomenclature', *Am. J. Psychiatry* 130, 1214–16.
Stebbins, G. L. (1950), *Variation and Evolution in Plants*, Columbia University Press, New York.
Stebbins, G. L. (1977), 'In defense of evolution: tautology or theory?' *Am. Nat.* 111, 386–90.

Stein, Z., Susser, M. and Guterman, A. V. (1973), 'Screening programme for prevention of Down's syndrome', *Lancet* 1, 305—10.

Stengel, E. (1974), *Suicide and Attempted Suicide*, Aronson, New York.

Stich, S. P. (1979), 'The recombinant DNA debate: some philosophical considerations', in D. A. Jackson and S. Stich (eds.), *The Recombinant DNA Debate*, Prentice-Hall, Englewood Cliffs, pp. 183—202.

Stoller, R. (1968), *Sex and Gender: On the Development of Masculinity and Femininity*, Science House, New York.

Stoller, R. (1973), 'Should homosexuality be in the APA nomenclature?' *Am. J. Psychiatry* 130, 1207—16.

Stoller, R. (1976), *Sex and Gender*, Volume II: *The Transsexual Experiment*, Aronson, New York.

Strickberger, M. W. (1975), *Genetics*, 2nd edn., Macmillan, New York.

Strouse, J. (1974), *Women and Analysis*, Grossman, New York.

Sturtevant, A. H. (1925), 'The effects of unequal crossing over at the bar locus in Drosophila', *Genetics* 10, 117—47.

Sturtevant, A. H. (1966), *A History of Genetics*, Harper, New York.

Suppe, F. (1974), *The Structure of Scientific Theories*, University of Illinois Press, Urbana.

Symons, D. (1979), *The Evolution of Human Sexuality*, Oxford University Press, New York.

Szasz, T. S. (1961), *The Myth of Mental Illness*, Delta, New York.

Thoday, J. M. and Gibson, J. B. (1962), 'Isolation by disruptive selection', *Nature* 193, 1164—6.

Tiger, L. and Shepher, J. (1975), *Women in the Kibbutz*, Harcourt, Brace, Jovanovich, New York.

Trivers, R. L. (1971), 'The evolution of reciprocal altruism', *Quart. Rev. Biol.* 46, 35—57.

Trivers, R. L. (1972), 'Parental investment and sexual selection', in B. Campbell (ed.), *Sexual Section and the Descent of Man, 1871—1971*, Aldine, Chicago.

Trivers, R. L. (1974), 'Parent-offspring conflict', *Am. Zool.* 14, 249—64.

Trivers, R. L. and Hare, H. (1976), 'Haplodiploidy and the evolution of social insects', *Science* 191, 249—63.

Trivers, R. L. and Willard, D. E. (1973), 'Natural selection of parental ability to vary the sex ratio of offspring', *Science* 179, 90—2.

Turnbull, A. C. (1978), 'Complications of amniocentesis', in J. Scrimgeour (ed.), *Towards the Prevention of Fetal Malformation*, Edinburgh University Press, Edinburgh, pp. 152—6.

U. S. Commission on Obscenity and Pornography (1970), *The Report of the U.S. Commission on Obscenity and Pornography*, Random House, New York.

Valenti, L. (1978), 'Fetal blood sampling in early pregnancy', in J. Scrimgeour (ed.), *Towards the Prevention of Fetal Malformation*, Edinburgh University Press, Edinburgh, pp. 273—86.

Valentine, J. W. (1978), 'The evolution of multicellular plants and animals', *Sci. Am.*, September, 140—58.

Valentine, R. C. (1978), 'Genetic engineering in agriculture', in J. Richards (ed.), *Recombinant DNA*, Academic Press, New York, pp. 59—70.

van den Berghe, P. (1979), *Human Family Systems: An Evolutionary View*, Elsevier, New York.

Veatch, R. M. (1974), 'Ethical issues in genetics', *Prog. Med. Gen.* 10, 223–64.

Waddington, C. H. (1957), *The Strategy of the Genes*, Allen and Unwin, London.

Wade, N. (1976), 'Inequality the main cause of world hunger', *Science* 194, 1142.

Wade, N. (1978a), 'New rulebook for gene splicers faces one more test', *Science* 201, 600–1.

Wade, N. (1978b), 'Cattle virus escapes from a P4 lab', *Science* 202, 290.

Wade, N. (1979a), 'Major relaxation in DNA rules', *Science* 205, 1238.

Wade, N. (1979b), 'Recombinant DNA: Warming up for big payoff', *Science* 206, 663–5.

Wade, N. (1979c), 'Supreme Court to say if life is patentable', *Science* 206, 664.

Wade, N. (1980), 'DNA: Chapter of accidents at San Diego', *Science* 209, 1101–2.

Wald, G. (1976), 'The case against genetic engineering', *The Sciences*, Sept./Oct. Reprinted in D. A. Jackson and S. Stich (eds.), *The Recombinant DNA Debate*, Prentice-Hall, Englewood Cliffs (1979), pp. 127–33.

Washburn, S. L. (1978), 'The evolution of man', *Sci. Am.*, September, 194–208.

Watson, J. D. (1968), *The Double Helix*, Atheneum, New York.

Watson, J. D. (1975), *Molecular Biology of the Gene*, 3rd edn., Benjamin, Menlo Park, Calif.

Watson, J. D. and Crick, F. H. C. (1953), 'Molecular structure of nucleic acids', *Nature* 171, 737–8.

Webb, T. *et al.* (1980), 'Amniocentesis in the West Midlands: report on 1000 births', *J. Med. Gen.* 17, 81–6.

Weinrich, J. D. (1976), 'Human reproductive strategy: I. Environmental Predictability and Reproductive Strategy; Effects of Social Class and Race. II. Homosexuality and Non-Reproduction; Some Evolutionary Models, Unpublished PhD. thesis, Harvard University.

Weinrich, J. D. (1978), 'Nonreproduction, homosexuality, transsexualism, and intelligence: I. A systematic literature search', *J. Homosexuality* 3 (3), 275–89.

Weller, J. M. (1969), *The Course of Evolution*, McGraw-Hill, New York.

West, D. J. (1967), *Homosexuality*, Penguin, Harmondsworth.

West Eberhard, M. J. (1975), 'The evolution of social behavior by kin section', *Quart. Rev. Biol.* 50, 1–33.

Westhoff, C. F. and Rindfuss, R. R. (1974), 'Sex preselection in the United States: Some implications', *Science* 184, 633–6.

Whewell, W. (1840), *Philosophy of the Inductive Sciences*, Parker, London.

Whitcomb, J. C. and Morris, H. M. (1961), *The Genesis Flood*, Presbyterian and Reformed Publishing Co., Nutley, N. J.

Whitehouse, H. K. K. (1965), *Towards an Understanding of the Mechanism of Heredity*, St. Martin's Press, New York.

Wiley, E. O. (1975), 'Karl R. Popper, systematics, and classification: a reply to Walter Boch and other evolutionary taxonomists', *Systematic Zool.* 24, 233–43.

Williams, G. C. (1966), *Adaptation and Natural Selection: A Critique of Some Current Evolutionary Thought*, Princeton University Press, Princeton.

Williams, M. B. (1970), 'Deducing the consequences of evolution: a mathematical model', *J. Theor. Biol.* 29, 343–385.

Williams, M. B. (1978), 'Ethical theories underlying the recombinant DNA controversy', in J. Richards (ed.), *Recombinant DNA*, Academic Press, New York, pp. 177–92.

Wilson, E. O. (1975a), *Sociobiology: The New Synthesis*, Belknap, Cambridge, Mass.

Wilson, E. O. (1975b), 'Human decency is animal', *The New York Times Magazine*, 12, October, pp. 38–50.

Wilson, E. O. (1978), *On Human Nature*, Harvard University Press, Cambridge, Mass.

Woodfield, A. (1976), *Teleology*, Cambridge University Press, Cambridge.

Wright, L. (1976), *Teleological Explanations*, University of California Press, Berkeley.

Wright, S. (1931), 'Evolution in Mendelian populations', *Genetics* 16, 97–159.

Yoshida, R. (1977), *Reduction in the Physical Sciences*, Dalhousie University Press, Halifax, N. S.

Zander, A. (1979), 'The discussion of recombinant DNA at the University of Michigan', in D. A. Jackson and S. Stich (eds.), *The Recombinant DNA Debate*, Prentice-Hall, Englewood Cliffs, pp. 39–55.

SUBJECT INDEX

NAME INDEX

Abrecht, P. 131
Adams, M. S. 213
Adkins, E. 240
Alexander, R. D. 200, 203, 208, 214
Allen, E. 99, 201–14, 224, 232–41, see also SftP
Alvarez, L. W. 30
Aquinas, T. 205
Ayala, F. J. 2, 10, 13, 15, 20, 23, 29, 33–4, 36, 38, 56–7, 67, 103, 111, 126–8, 193

Balian, R. 244
Bandura, A. 229
Barash, D. 191
Bardwick, J. 223
Barker, A. D. 28, 72
Barlow, G. W. 191
Bartsch, F. 135, 141, 145
Beatty, J. xviii, 4, 15, 89
Beck, L. W. 132
Beckner, M. 26
Bell, A. 210, 241, 249–56, 264
Benzer, S. 116–17
Berg, P. 159
Bethell, T. 3, 28, 36
Bieber, I. 154, 245, 258–91
Birch, C. 131
Black, M. 101
Blumberg, B. 145
Boorse, C. xviii, 97, 247, 253–4, 258, 261–3, 265–7, 269
Borgaonkar, D. 133
Bowler, P. J. 43
Bowser-Riley, S. 141
Braithwaite, R. B. 5
Brewster, D. 89
Brock, D. 135, 142–3, 151
Brooks-Gunn, J. 229
Brown, A. E. 86

Brown, R. 249
Bunge, M. xviii
Burchfield, J. D. 54
Burkhardt, R. 46

Cain, A. xviii, 27
Callahan, D. 161
Campbell, D. T. 95
Cannon, H. G. 46
Caplan, A. xviii, 72, 191
Carlson, E. O. 13, 114, 116
Carter, F. 149
Carter, G. S. 20
Chakrabarty, A. 169–70
Chambers, R. 50
Chargaff, E. 158–9, 179–80, 184
Childs, B. 135
Churchill, F. B. 52
Churchill, W. 260
Cohen, C. 161
Cole, J. R. 65, 160, 220
Cole, S. 65, 160, 220
Copernicus, N. 1, 83, 107, 238
Cove, A. 135, 149
Cracraft, J. 67
Crick, F. 11, 102, 117–18
Crow, J. F. 141
Curtis, R. 176

Darden, L. xviii, 244
Darlington, C. 52
Darwin, C. R. 1, 15, 21, 23–4, 28–31, 44, 46, 50, 52–3, 63–5, 81–2, 84, 90–4, 102, 118, 121, 130, 183, 192, 196, 221–2, 238
Davis, B. 160, 179–82
Dawkins, R. 16, 23, 191, 203, 233
De Fries, J. 212
Dickerson, R. 42
DHEW, 160

295